Physical Chemistry of Non-aqueous Solutions of Cellulose and Its Derivatives

Wiley Series in Solution Chemistry

Volume 1
pH and Buffer Theory—A New Approach
H Rilbe
Chalmers University of Technology, Gothenburg, Sweden

Volume 2
Octanol–Water Partition Coefficients: Fundamentals and Physical Chemistry
J Sangster
Sangster Research Laboratories, Montreal, Canada

Volume 3
Crystallization Processes
Edited by
H Ohtaki
Ritsumeikan University, Kusatsu, Japan

Volume 4
The Properties of Solvents
Y Marcus
The Hebrew University of Jerusalem, Israel

Volume 5
Physical Chemistry of Non-aqueous Solutions of Cellulose and Its Derivatives
V V Myasoedova
Institute of Chemical Physics, Russian Academy of Sciences, Moscow, Russia

Physical Chemistry of Non-aqueous Solutions of Cellulose and Its Derivatives

Vera V. Myasoedova

Institute of Chemical Physics, Russian Academy of Sciences, Moscow, Russia

Wiley Series in Solution Chemistry
Volume 5

JOHN WILEY & SONS
Chichester · New York · Weinheim · Brisbane · Singapore · Toronto

Baffins Lane, Chichester,
West Sussex PO19 1UD, England

National 01243 779777
International (+44) 1243 779777
e-mail (for orders and customer service enquiries): cs-books@wiley.co.uk
Visit our Home Page on http://www.wiley.co.uk
or http://www.wiley.com

Other Wiley Editorial Offices

John Wiley & Sons, Inc., 605 Third Avenue,
New York, NY 10158-0012, USA

WILEY-VCH Verlag GmbH, Pappelallee 3,
D-69469 Weinheim, Germany

Jacaranda Wiley Ltd, 33 Park Road, Milton,
Queensland 4064, Australia

John Wiley & Sons (Asia) Pte Ltd, 2 Clementi Loop #02-01,
Jin Xing Distripark, Singapore 129809

John Wiley & Sons (Canada) Ltd, 22 Worcester Road,
Rexdale, Ontario M9W 1L1, Canada

Library of Congress Cataloging-in-Publication Data

Myasoedova, Vera V.
Physical chemistry of non-aqueous solutions of cellulose and its derivatives / by Vera V. Myasoedova.
p. cm. — (Wiley series in solution chemistry; v. 5)
Includes bibliographical references and index.
ISBN 0-471-95924-3 (alk. paper)
1. Cellulose. 2. Solution (Chemistry) I. Title. II. Series
QD323 . M93 2000 99–040221
547'.782 21—dc21

British Library Cataloguing in Publication Data

A catalogue record for this book is available from the British Library

ISBN 0 471 95924 3

Typeset in 10/12pt Times by Thomson Press (India) Ltd, New Delhi
Printed and bound in Great Britain by Biddles Ltd, Guildford and King's Lynn
This book is printed on acid-free paper responsibly manufactured from sustainable forestry, in which at least two trees are planted for each one used for paper production.

Contents

Series Preface

There are many aspects of solution chemistry. This is apparent from the wide range of topics which have been discussed during recent International Conferences on Solution Chemistry and International Symposia on Solubility Phenomena. The Wiley Series in Solution Chemistry was launched to fill the need to present authoritative, comprehensive and up-to-date accounts of these many aspects. Internationally recognized experts from research or teaching institutions in various countries have been invited to contribute to the Series.

Volumes in print or in preparation cover experimental investigation, theoretical interpretation and prediction of physical chemical properties and behaviour of solutions. They also contain accounts of industrial applications and environmental consequences of properties of solutions.

Subject areas for the Series include: solutions of electrolytes, liquid mixtures, chemical equilibria in solution, acid–base equilibria, vapour–liquid equilibria, liquid–liquid equilibria, solid–liquid equilibria, equilibria in analytical chemistry, dissolution of gases in liquids, dissolution and precipitation, solubility in cryogenic solvents, molten salt systems, solubility measurement techniques, solid solutions, reactions within the solid phase, ion transport reactions away from the interface (i.e. in homogeneous, bulk systems), liquid crystalline systems, solutions of macrocyclic compounds (including macrocyclic electrolytes), polymer systems, molecular dynamic simulations, structural chemistry of liquids and solutions, predictive techniques for properties of solutions, complex and multi-component solutions applications, of solution chemistry to materials and metallurgy (oxide solutions, alloys, mattes etc.), medical aspects of solubility, and environmental issues involving solution phenomena and homogeneous component phenomena.

Current and future volumes in the Series include both single-authored and multi-authored research monographs and reference level works as well as edited collections of themed reviews and articles. They all contain comprehensive bibliographies.

Volumes in the Series are important reading for chemists, physicists, chemical engineers and technologists as well as environmental scientists in academic and industrial institutions.

May 1996 Peter Fogg

Preface

Cellulose is the most abundant organic polymer on earth and is the subject of countless publications. However, there has been no systematic account of the solution chemistry of cellulose and its derivatives. In solution, cellulose derivatives can form liquid crystals which change to a solid state with unique optical and physico-mechanical properties. This book explains how experimental data and computer simulation can give an insight into the factors which influence the interaction of solvent and solute. It is also shown how phase transitions in solution can be predicted from the solvency of non-aqueous solvents for cellulose and its derivatives. Methods of obtaining thermodynamic parameters for solvation in non-aqueous solvents are explained. Of special interest are the rheological properties of lyotropic liquid crystals based on cellulose derivatives.

The book is intended to stimulate the development of the potentialities of the optical properties of solutions of cellulose and its derivatives in non-aqueous solvents.

I am greatly indebted to Dr Peter Fogg for editorial assistance with the text.

Vera V. Myasoedova
Moscow
August 1999

INTRODUCTION

Polysaccharides occur widely as fibrous constituents of plant cell walls. Payen, in 1842, gave the name cellulose to the material which is completely hydrolysed by acid to give exclusively glucose. On this basis cotton cellulose is almost pure cellulose. The term cellulose is also used more loosely in a technological context to mean the residues obtained when materials of plant origin are subjected to certain pulping processes. These pulps consist mainly of cellulose in the strict sense but may contain various proportions of non-cellulosic components. In this book the term cellulose will be used in the sense intended by Payen.

Cellulose consists of long chains of glucopyranose units linked at the 1 and 4 positions through glucosidic bonds (Figure 1). Alternate glucopyranose units are rotated through about 180°. The number of glucopyranose units can range from about 2000 to about 25 000. The degree of polymerization depends on the source and the history of the sample. Samples are polydisperse. The structure of the chain may be represented as in Figure 2. The chains are stiffened by van der Waals forces and by inter- and intramolecular hydrogen bonding. Single chains never exist under natural conditions but occur in the form of microfibrils which consist of many ordered parallel chains.

Cellulose can exist in more than one crystalline form with different orientations of parallel chains relative to each other. The most common crystalline form in nature, cellulose I, is metastable. Dissolution and reprecipitation lead to the stable form, cellulose II. This form is manufactured commercially and sold as Rayon.

Cellulose molecules contain three hydroxy groups per link, except in the case of the end links. These hydroxyl groups can be partially or completely converted to ester or ether linkages. Hydrolysis under acid conditions leads to a progressive breaking of the glucosidic bonds and ultimately to the formation of glucose. Oxidation converts hydroxyl groups to aldehydo or to carboxyl groups and tends to cause cleavage of glycosidic linkages, leading to shorter chains. A variety of oxidation products are formed depending on the conditions. Cellulose forms a range of simple or complex addition compounds, for example with organic bases, metal salts and cuprammonium hydroxide. Cellulose itself dissolves in a limited range of solvents to form solutions. The esters dissolve in a wider range of solvents.

Cellulose, in its various forms and degrees of purity, is readily available and cheap. The ease in which it can be converted to derivatives make it an attractive raw

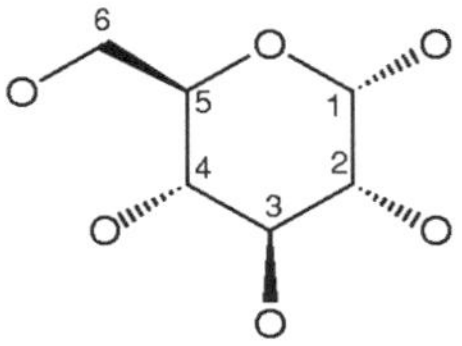

Figure 1 The glucopyranose ring

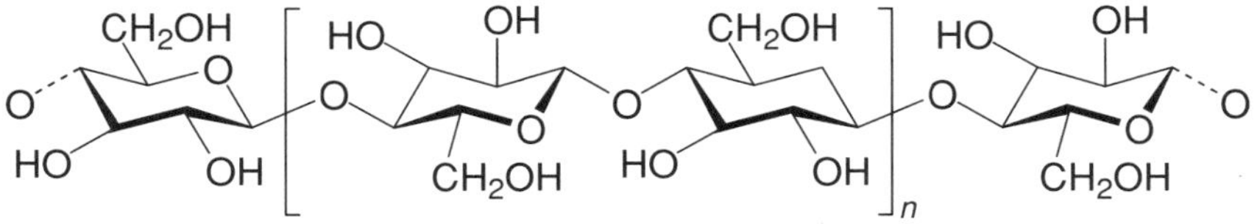

Figure 2 The structure of the cellulose chain. Reprinted with premission from R.M. Brown, Jr, M. Saxena and K. Kudlika, *Trends Plant Sci.* **1996**, *1*, 149–156. Copyright 1996 Elsevier Science

material. Much has been written about the behaviour of solid cellulose but rather less about the interesting behaviour of cellulose and cellulose derivatives in the liquid phase. This book is intended to give an overview of the various approaches which have been made to investigate the structure and the properties of solutions of cellulose and its derivatives. Some of these properties have significant industrial potential.

CHAPTER 1

Phase Equilibria and Liquid Crystalline Order in Solutions of Cellulose and Its Derivatives

1 PREDICTION OF POLYMER SOLUBILITY AND FORMATION OF LYOTROPIC LIQUID CRYSTALS IN SOLUTIONS OF CELLULOSE AND ITS DERIVATIVES

The selection of effective non-aqueous solvents for cellulose is of great importance. The development of a theoretical basis for the selection of solvents and prediction of the solubility in non-aqueous solvents of cellulose, its ethers and esters is based on agreement between theoretical calculations and experimental measurements of physico-chemical properties of solutions.

Prediction and control of the behaviour of polymeric systems when the temperature and concentrations are changed can only be achieved if there are complete diagrams of state over wide ranges of composition and temperature. Obtaining diagrams of state in such systems is complicated by the difficulty in achieving equilibrium and by the characteristic manifestations of gel formation in the region of high polymer concentrations [1, 2], and also by manifestations of anisotropic properties. The difficulties mentioned are largely due to the peculiarities of the chemical and supramolecular structure of cellulose. These are considered in detail in monographs dealing with the subjects [3–5].

Extensive reviews have been published which give details of individual and composite organic solvents for cellulose [6–12]. There are also a series of patents [12–20] covering the use of ternary amide *N*-oxides, mixtures of dimethylacetamide with lithium chloride, trifluoroacetic acid with chlorinated hydrocarbons and of some other systems which have a practical application.

The application of one of the basic criteria of solubility—the approximate equality between the Hildebrand solubility parameters for polymer (δ_p) and solvent (δ_s) [21]—has a number of limitations. Even if δ_p is not approximately equal to δ_s there will be appreciable solubility in $\sim 50\%$ of cases. Not only should the chemical structure of polysaccharides be taken into account but also their supramolecular structure. Globular supramolecular structure with the size of globules up to several hundreds of ångstroms in diameter is, as a rule, observed in the case of amorphous polymers [22, 23]. The forces of surface and inter-phase tension of solvent and polymer, taking into account the curvature of the surface on which they act, can be used to determine the conditions of polymer solubility. This approach has been developed by Askadskii and Matveyev [22].

In more recent years, various attempts have been made to develop a more precise criterion for the prediction of solubility. One method takes into account the change in entropy when a polymer is dissolved and is based on a precise estimation of the effect of hydrogen bonds [24]. This method is based on the classical concepts of Hildebrand and Scott and on Hansen's equation [25]:

$$\frac{\$_{\mathrm{m}}}{C} = R_A^2 - 4({}^{\mathrm{s}}\delta_{\mathrm{d}} - {}^{\mathrm{p}}\delta_{\mathrm{d}})^2 - ({}^{\mathrm{s}}\delta_{\mathrm{pol}} - {}^{\mathrm{p}}\delta_{\mathrm{pol}})^2 - ({}^{\mathrm{s}}\delta_{\mathrm{H\text{-}bond}} - {}^{\mathrm{p}}\delta_{\mathrm{H\text{-}bond}})^2 \tag{1.1}$$

where $\$_{\mathrm{m}}$ = dissolving power of solvent for polymer; C = a constant; R_{A} = radius of the solvation sphere; ${}^{\mathrm{s}}\delta_{\mathrm{d}}$, ${}^{\mathrm{p}}\delta_{\mathrm{d}}$ = Hansen dispersion parameters; ${}^{\mathrm{s}}\delta_{\mathrm{pol}}$, ${}^{\mathrm{p}}\delta_{\mathrm{pol}}$ = polar parameters; and ${}^{\mathrm{s}}\delta_{\mathrm{H\text{-}bond}}$, ${}^{\mathrm{p}}\delta_{\mathrm{H\text{-}bond}}$ = hydrogen bonding parameters. Superscript s refers to the solvent and p to the polymer.

In order to take the effect of H-bonds and the entropy factor into account more precisely, it has been suggested that the dependence of $\$_{\mathrm{m}}$ (solvent solvency for polymer) should include the following: the molar volume of solvent, V_{s}, the molar volume of the repeating link of the polymer, V_{p}, the so-called 'structural factor,' b_{s}, of the solvent which relates to the presence or absence of H-bonded chains, and various solubility parameters (dispersion, dipole–dipole interactions, polymer–solvent H-bonds, etc.).

The complete equation [see Eq. (1.11)] also includes some variable parameters that require experimental measurement: a modified solubility parameter for the solvent, ${}^{\mathrm{s}}\delta'$, determined from the solubility of solid alkanes in the solvent; a modified solubility parameter for the polymer, ${}^{\mathrm{p}}\delta'$; a constant A, characteristic of the solvent; and a stability constant, K, for the formation of hydrogen bonds between the polymer and the main functional groups of the solvent.

From the thermodynamic point of view, the ability of polymer to exist in a particular phase is expressed by its chemical potential μ_{sub} in that phase. This ability is high if the chemical potential has a low value. The solvent solvency for a substance 'sub' can be quantitatively expressed as

$$\$^0_{\mathrm{sub}} = \frac{\mu'_{\mathrm{sub}} - \mu_{\mathrm{sub\ solut}}}{RT} \tag{1.2}$$

where R is the gas constant (R = 8.315 J mol^{-1} K^{-1}) and T is the absolute temperature. Equation (1.2) can be written as follows:

$$\$^0_{\mathrm{sub}} = \frac{-\Delta H_{\mathrm{transit}}}{RT} + \frac{-\Delta S_{\mathrm{transit}}}{R} \tag{1.3}$$

The first contribution, $\Delta H_{transit}$, is the enthalpy change during the transition of 1 mol of substance 'sub' from its own phase into solution. The second contribution is the corresponding change in entropy.

If substance 'sub' is a polymer with r repeating segments, solvency can be represented as

$$\$^0_{sub} = r\$_{seg} + \text{ contribution of end groups} \tag{1.4}$$

where $\$_{seg}$ is the disolving power of the solvent for individual segments given by

$$\$_{seg} = \frac{\mu'_{seg} - \mu_{seg\ solut}}{RT} \tag{1.5}$$

where $\mu'_{seg\ solut}$ and μ'_{seg} are the chemical potentials of the repeating segments in solution and in their own phase, respectively. It is noteworthy that $\$_{seg}$, characterizing the solvency of a given solvent for a segment of a given polymer, is different for different polymers in the same class. For example, it will decrease when an amorphous polymer becomes semicrystalline. It also depends on the molecular mass: the value of \$ increases with decreasing molecular mass.

When the degree of polymerization is high (high r value), the action of the end groups can be neglected, and $\$_p$ is proportional to $\$_{seg}$. Consequently, the polymer is soluble up to the mole fraction ϕ_p in the solvent if the value of solvency $\$_{seg}$ is positive or equal to zero.

When the degree of polymerization is low, as for oligomers, the contribution of the end groups to the solvency should not be neglected.

For a polymer which forms H-bonds with the solvent, the solvency can be calculated by the equation

$$\$_{seg} = -A - D + B + H \tag{1.6}$$

assuming that

$$D = \frac{\bar{V}_{seg}}{RT}(1 - \phi_p)^2({}^s\delta' - {}^p\delta')^2 \tag{1.7}$$

$$B = (0.5 + b_{solut})(1 - \phi_p)\frac{\bar{V}_{seg}}{\bar{V}_{solut}} \tag{1.8}$$

$$H = \ln\left(1 + K\frac{1 - \phi_p}{\bar{V}_{solut}}\right) \tag{1.9}$$

where K ($cm^3\,mol^{-1}$) is the stability constant of polymer–solvent hydrogen bond formation.

Hydrogen bonds between the segments occur when these include OH or NH groups. For the calculations of solvency taking into account polymer–polymer hydrogen bonds there is an additional contribution, C, in Eq. (1.6) which is approximated well by the expression

$$C = \ln\left(\frac{1 + K_{pp}/\bar{V}_{seg}}{1 + K_{pp}\phi/\bar{V}_{seg}}\right) \tag{1.10}$$

where K_{pp} ($cm^3\,mol^{-1}$) is the stability constant for segment–segment H-bond formation.

The complete equation is as follows:

$$\begin{aligned}\$_{seg} &= -A - \frac{\bar{V}_{seg}}{RT}({}^{s}\delta' - {}^{p}\delta')^2(1-\phi_p)^2 + (0.5 + b_{solut})\frac{\bar{V}_{seg}}{\bar{V}_{solut}}(1-\phi_p) \\ &+ \ln\left[1 + K(1-\phi_p)/\bar{V}_{solut}\right] - \ln\left[(1 + K_{pp}/\bar{V}_{seg})/(1 + K_{pp}\phi_p/\bar{V}_{seg})\right]\end{aligned} \tag{1.11}$$

Huyskens and Haulait-Pirson [24] were able to predict precisely the solvency of three polymers, poly(vinyl acetate), poly(methyl methacrylate) and poly(ethyl methacrylate), in 25 solvents and their binary mixtures in more than 95% of cases using Eq. (1.11) and only in 78% of cases using Hansen's equation.

Myasoedova and co-workers [26, 27] have put forward an equation for the parameter of interaction of cellulose and its derivatives with non-aqueous solvents, χ, taking into account non-polar and dispersion interactions and H-bonding, i.e.

$$\chi = \chi_{solv} + \frac{\bar{V}_1}{RT}\left\{\delta_p - \delta_s\left[\phi(\delta_{pol}, \delta_d, \delta_{H\text{-bond}})\right]\right\} \tag{1.12}$$

where χ_{solv} is the Flory parameter for the solvent and $\bar{V}_1$ the partial molar volume of the solvent. This equation has proved to be more suitable than earlier equations for dealing with systems involving cellulose and its derivatives.

Nesterov and Lipatov published numerous phase diagrams for various polymer–solvent, polymer–solvent–solvent and polymer–polymer–solvent systems [28]. In addition, Scott [29] calculated the general form of phase diagrams for various systems showing the miscibility of components: polymer–solvent (1)–solvent (2).

Non-traditional approaches to the problem of simulation of the phase equilibria in binary and multicomponent systems were developed by Papkov and Kulichikhin [30]. Calculations of phase diagrams for athermal solutions of polymers were published by Birshtein and Merkuryeva [31] and by Valenti and Sartirana [32] (Figures 1.1 and 1.2).

Prediction of the phase state in multicomponent polymer systems and of phase equilibrium in solutions of rigid-chain macromolecules [33] is more difficult than for other polymer systems.

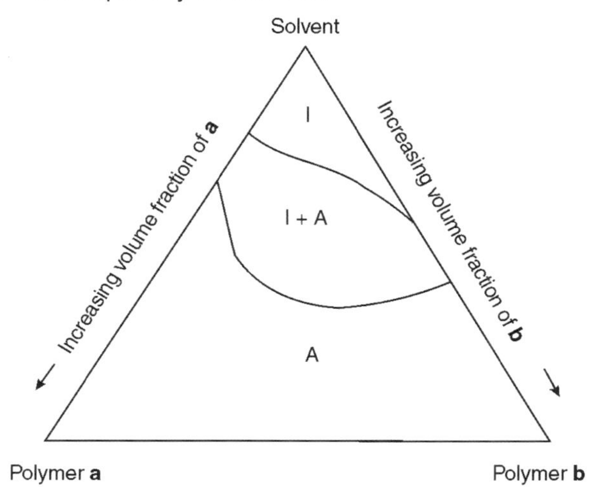

Figure 1.1 Ternary phase diagram for two rod like polymers ($X_a = 40$, $X_b = 20$) in pure solvent where X_a and X_b are the degrees of polymerization of polymers **a** and **b**, respectively

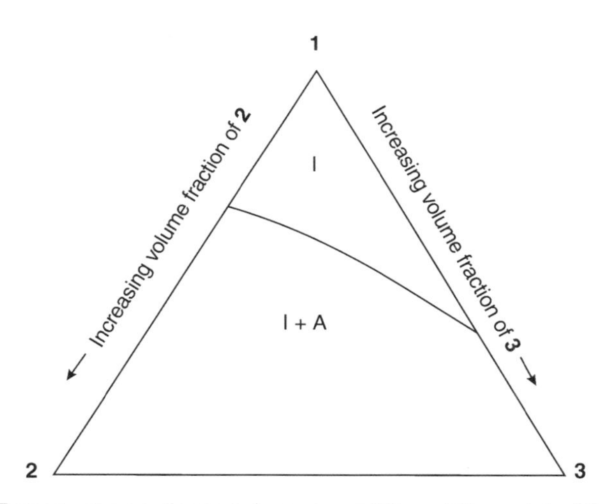

Figure 1.2 Ternary phase diagram for solvent (1), rod like solute (2) and statistical tangle (3). $X_2 = X_3 = 20$

Of particular importance are papers dealing with the behaviour of polymers for which the mesomorphic properties relate to the skeletal rigidity or partial skeletal rigidity [32]. Such papers are based on theories of phase equilibria for low- and high-molecular-mass mesogens and include lattice theory and the virial approximation. Asymmetry of molecular form is the main criterion for liquid-crystalline order common to all mesogens. The discovery that fibres may have increased tensile strength if they are formed in solutions of nematics has excited interest in studies of rigid-chain and semi-rigid-chain polymers. Theoretical studies of polymeric liquid crystals, which have been extended during recent years, have involved special attention to the significance of the axial ratios of molecules. Systems of rigid rod-like particles, that are distributed within a solvent, can be described by the lattice model. Flory [34] used this model to establish the interrelation between the free energy of mixing of solvent and polymer and the axial ratio:

$$\Delta G_{mix}/kT = n_1 \ln V_1 + n_2 \ln V_2 - (n_1 + yn_2) \times \ln[1 - V_2(1 - y/x)] - n_2[\ln(xy^2) - y + 1] + \chi x n_2 V_1 \quad (1.13)$$

The dissolved molecules were represented as rigid rods, for which interaction consisted only of repulsion on collision. Irregularity in the orientation of the rods relative to the axes of domains was expressed by thermodynamic parameters. In Eq. (1.13) n_1 and n_2 are the number of solvent and polymer molecules, respectively, V_1 and V_2 are the volume fractions of solvent and polymer, respectively, χ is the interaction parameter, which is proportional to the reciprocal of temperature x is the ratio of length to diameter of the rods and y is an index of disorientation varying from one for perfect order to x for complete disorder.

In the case of irregularity in the orientation, Eq. (1.13) becomes similar to a well known Flory–Huggins expression. The rigidity of Flory's chains is characterized by the degree of bending f. At $f > 0.63$, the polymer is able to exist in a liquid-crystalline (LC) state; at lower values it is not. The value of f is related to the degree of flexibility, $\langle r^2 \rangle / r_{max}$. The relationship which is usually used for the flexible-chain polymers is

$$\langle r^2 \rangle / r_{max} = l(2 - f)/f \quad (1.14)$$

It follows that the critical expression for formation of the LC phase is

$$\langle r^2 \rangle / l r_{max} > 2.16$$

where $\langle r^2 \rangle$ is the average-square distance between the chain ends, l is the length of a unitary link and r_{max} is the contour length of the chain.

On the basis of statistical thermodynamics, Flory proposed a quasi-crystalline lattice model for suspensions of rigid rods in a medium of low-molecular-mass liquid. This model made use of the approximation of a self-consistent field. Calculations showed that at a certain volume fraction, V_2^*, regular packing of rods becomes thermodynamically favourable and a first-order phase transition from an isotropic to an anisotropic system should occur.

The diagram proposed by Flory [34] to describe the phase state of a system 'rigid-chain polymer–liquid' is characterized by a change in χ due to temperature change or due to introduction into the system of a second low-molecular-mass liquid. This addition transforms a 'bad' solvent into a 'good' one.

A critical analysis of the lattice theory, taking into account the shortcomings of its predictive ability, has been published by Nies *et al.* [35]. They pointed out the impossibility of taking excess volume into account. The following equation was published by Nies *et al.* [35] but was proposed nearly simultaneously but independently by Flory, Huggins, Staverman and Van Santin:

$$\Delta G_{\mathrm{mix}}/RT = n_1 \ln \phi_1 + n_2 \ln \phi_2 + \Gamma \tag{1.15}$$

where $\phi_1 = n_1/N, \phi_2 = n_2 m_2/N, N = n_1 + n_2 m_2$ and m_2 is the number of lattice sites which are occupied by a polymer chain. The value of Γ is given by

$$\Gamma/N = g\phi_1\phi_2 = (g_{\mathrm{s}} + g_{\mathrm{h}}/T)\phi_1\phi_2 \tag{1.16}$$

where g is the interchange energy, g_{s} is the entropic contribution and g_{h} is the enthalpic term.

Factors influencing the interaction function were considered. The dependence of the entropy of mixing on the chain length was estimated as follows:

$$\Delta G/(NRT) = \phi_1 \ln \phi_1 + (\phi_2/m_2) \ln \phi_2 + \Gamma/N \tag{1.17}$$

The contribution Γ is for correction of the first two combinatorial terms. This is necessary for agreement with the real thermodynamic properties. ϕ_1 and ϕ_2 are the mole fractions of solvent and polymer. The 'relative chain length,' m_2, is given by

$$m_2 = \bar{V}_2 M_2/\bar{V}_1 \tag{1.18}$$

where $\bar{V}_2$ and $\bar{V}_1$ are the partial molar volume of the polymer in the liquid state and partial molar volume of the solvent, respectively. The second osmotic coefficient α_2 has been shown to depend on molecular weight and to be related to the Flory interaction parameter, i.e.

$$\alpha_2 \sim (^1\!/_2 - \chi) \tag{1.18a}$$

$$\chi = (g - \phi_1 \partial g/\partial \phi_2)_{\phi_2=0} \tag{1.19}$$

For a dilute polymer solution,

$$\alpha_2 \sim (^1\!/_2 - \chi)h(Z) \tag{1.19a}$$

where $h(Z)$ is a function of the excluded volume, which includes the dependence on the chain length in quantitative form. The temperature dependence of the interaction function is given.The heat capacity of a liquid under constant pressure depends on temperature. The value of ΔC_p, the change in heat capacity on mixing, also changes with concentration and is given by

$$\Delta C_p = (C_0 + C_1 T/K)\phi_1\phi_2 \tag{1.20}$$

where C_0 is a constant and C_1 the concentration.

Combination of this equation with the following two equations enables the change in interaction function with change in temperature to be estimated:

$$\Delta H = \int \Delta C_p \mathrm{d}T \qquad \Delta S = \int (\Delta C_p/T)\mathrm{d}T \tag{1.21}$$

(ΔH and ΔS refer to changes due to mixing). It follows that

$$g = g_a + g_b/(T/K) + g_c/(T/K) + g_d \ln(T/K) \tag{1.22}$$

where g_a and g_b are found from integration constants and $g_c = -C_1/2NR$ and $g_d = -C_0/NR$.

It is known that polymers such as polyisocyanides and polyorganophosphates and also highly soluble cellulose derivatives possess lyotropic mesomorphism. These polymers are characterized by a Mark–Houwink coefficient, α, where $0.6 < \alpha < 0.9$. This shows that they have less rigid chains than the lyotropic mesomorphous polymers that are approximated by the behaviour of rigid rods such as poly(*p*-benzamide) or poly(*p*-alkyl isocyanates). On the other hand, the persistent lengths and axial ratios indicate that these polymers have chains that differ from those of the rigid-chain polymers [36]. Thermodynamic principles indicate [37] that the formation of an LC phase is due to the change in the energy of polymer–solvent interaction favouring increased rigidity of the chain.

There is a correlation between V_2^* and 'solvent quality' and also a correlation between V_2^* and χ [38] in non-aqueous solutions of the cellulose derivatives. There have also been interesting studies on the mesomorphous behaviour of cellulose and its derivatives. Two rod-like or irregularly spiral polymers have been reported [39].

Splitting into two phases, isotropic and nematic, depends on the critical axial ratio, as mentioned earlier. This, in turn, depends on polymer concentration. The approximate correlation shows that there is an athermal limit corresponding to χ equal to zero. This corresponds to the minimum value of V_2^* at which the nematic phase is stable:

$$V_2^* = (8/\chi)(1 - 2/\chi) \tag{1.23}$$

Pawlowski *et al.* [40] have given examples of non-athermal solutions in which the regions of co-existence of the isotropic and anisotropic phases depend on the values of χ for different axial ratios. They state that there are limits to the double-phase

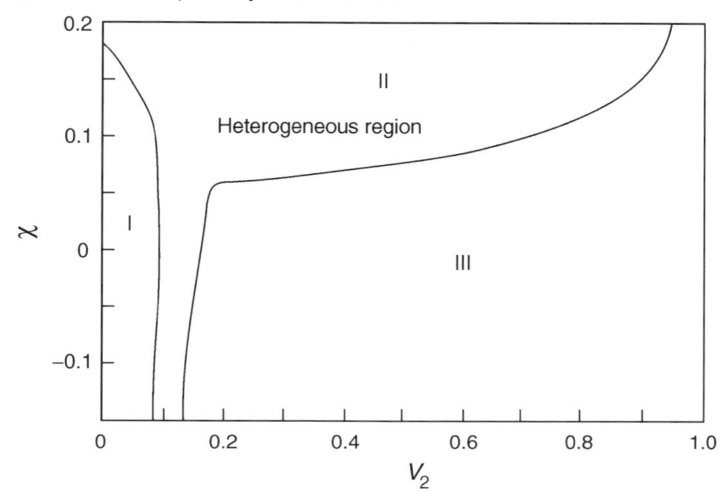

Figure 1.3 Predicted compositions of phases in equilibrium when the ratio of length to diameter of rod shaped molecules, X, is equal to 100. Values of the interaction parameter, χ, are plotted as ordinate against volume fraction of solute, V_2. Region I corresponds to isotropic solutions and region III to ordered phases. Region II corresponds to a heterogeneous system

region in which isotropic and anisotropic phases co-exist and that there exists a wide double-phase region where the anisotropic phase is very concentrated and the isotropic phase is dilute. Similar statements have been made by other authors. The transition from a narrow into a wide double-phase region occurs when the thermodynamic qualities of solvent become worse ($\chi_1 > 0$), which is characteristic of endothermal solutions.

Splitting into two phases was predicted by Flory for polymers with high molecular mass and small values of f (degree of flexibility). Consequently, the narrow region (Figure 1.3) displaces to high volume fractions and is much more extended [34].

In 1979, Flory and Ronca [41] found a more precise method of interpreting systems containing rigid rods. Later Matheson and Flory [42] considered systems of rigid impermeable rod-like molecules, that are subject to orientational effects. They took into account interactions between the pairs of segments in contact. Orientational interactions have also been discussed by Birshtein and Merkuryeva [37].

The temperature at which the nematic–isotropic transition in a heterogeneous liquid occurs is determined by the equation

$$\bar{T}_N = T_{N1}/XT^* \tag{1.24}$$

where T^* is the critical temperature of phase transition and X is the axial ratio of length to diameter of polymer chains. When $X > 6.417$, the nematic phase is stable at all temperatures, since the molecules are asymmetric [38]. The temperature of the transition to the ordered liquid-crystalline phase depends, to a large extent, on flexibility of molecules and is dependent on the change in flexibility with change of temperature [43]. Values of the parameter of order, **S**, for various values of chain flexibility and the dependence on temperature have been published by Bosch *et al.* [43, 44].

A linear dependence of the phase transition temperature on polymer mole fraction has been predicted from models. The slope is expected to increase with increasing chain rigidity. This linear relationship over a wide concentration region has been found to be characteristic of a series of systems [44]. Classification with respect to chain rigidity and analysis of the data for various cellulose derivatives have been carried out [45]. The marked change of slope in the phase boundary which is found in some cases is evidently due to the destruction of the nematic axes and the strengthening of inter-chain interactions or to specific polymer–solvent interaction and to the dependence of the chain rigidity on temperature [46].

New effects of chain flexibility have been found on the transition temperature and on the nematic parameter of order of a polymer. Bosch *et al.* [44] have proposed a theory of the average field for nematic–isotropic phase transition in semi-rigid chain polymers. This is based on the approximation which is applicable for long flexible chains for which the ratio between the persistent length and the total length of the entire chain is low. A possible configuration of a chain is given by a continuous space curve $r(s)$, where s is the length of the arc measured from the origin of the polymer and r is the direction of the tangent to the polymer chain at the point defined by s. It is found for short rigid chains that v, the effective monomer–chain interaction energy, increases with increase in the degree of polymerization. This is due to an increase in the degree of branching and greater orientational interaction. By analogy with the method used for the model of liquid membranes [44], it is possible to calculate the Helmholtz free energy A and partition function Z of the solutions of nematic polymers from the equations

$$A = kT \ln Z$$
$$Z = \int \mathrm{d}\boldsymbol{R}\,\mathrm{d}\boldsymbol{R}'\,G(\boldsymbol{R}\boldsymbol{R}'L) \tag{1.25}$$

where L is the length of a chain in monomer units and hence the degree of polymerization and $G(\boldsymbol{RR}'L)$ is a statistical weight function.

The parameter of order, **S**, may be estimated by the equation for flexible chains:

$$\boldsymbol{S} = \frac{1}{L}\int_0^L \mathrm{d}s\,\frac{1}{Z}\,\mathrm{d}\boldsymbol{R}\,\mathrm{d}\boldsymbol{R}'\,\mathrm{d}\boldsymbol{R}''\,G\left(\boldsymbol{R}\boldsymbol{R}'L - s\right)\left[\frac{3}{2}\cos^2\theta(s) - \frac{1}{2}\right]G(\boldsymbol{R}''\boldsymbol{R}'s) \tag{1.26}$$

Furthermore, it is possible to determine the change in the conformation of the polymer chain resulting from the nematic–isotropic transition. The calculation of the ratio of projection of vector $\boldsymbol{R}_0$ ends of the chain on the initial vector $\boldsymbol{R}'$ in the nematic and isotropic phases corresponds to the relationship

$$\langle \alpha \rangle = \frac{\langle \boldsymbol{R}'\boldsymbol{R}_0 \rangle_{\text{nematic}}}{\langle \boldsymbol{R}'\boldsymbol{R}_0 \rangle_{\text{isotrop}}} \tag{1.27}$$

The change into a more extended state of a molecule corresponds to a jump of $\langle \alpha \rangle$ and by the increase $\langle \alpha \rangle > 1$. The theory includes a special case of infinite rigid chains. As the rigidity increases, the bending energy increases to reach a constant value and $r(s)$ remains constant in space. Molecules then behave as simple rigid rods. In this case,

$$G(\boldsymbol{R}\boldsymbol{R}'L) = \delta(\boldsymbol{R} - \boldsymbol{R}')\exp[-\beta L \nu(L)] \tag{1.28}$$

where $\beta = 1/kT$.

The data for LC systems consisting of small molecules where $L = 1$ are reproduced by this equation. If the chains are flexible, then function $G(\boldsymbol{R}\boldsymbol{R}'L)$ is a solution to a Schrödinger equation:

$$\left(\frac{\partial}{\partial L} - \frac{1}{2\beta\kappa}\Delta\boldsymbol{R}' + V(L) - \frac{1}{2}VLS^2\right)G(\boldsymbol{R}\boldsymbol{R}'L) = \delta(L)\delta(\boldsymbol{R} - \boldsymbol{R}') \tag{1.29}$$

where κ is the bending elastic constant of a polymer chain.

Owing to the spherical symmetry, spherical harmonics can be used and the equation can be resolved quantitatively. Expressions for the chemical potentials and the double-phase equilibrium between the nematic and isotropic phases of the semi-rigid chain macromolecules were published by Gilbert [47]. The presence of very few flexible parts of the chain increases the disordering in the anisotropic phase and narrows the double-phase region. Gilbert gave various examples of systems with numerous phase transitions.

The centenary of the description of the first liquid-crystalline substance, cholesteryl benzoate, by the Austrian botanist F. Reisitzer was in 1988. The liquid-crystalline state is now know to be widespread among biological systems [48, 49]. Much information has been published on lyotropic and thermotropic liquid-crystalline polymers [50, 51]. General approaches to the phase equilibrium of the lyotropic liquid-crystalline systems have been reviewed [52, 53].

Experimental data for the region of phase equilibria for cellulose and other polymers with moderately rigid chains have been extremely limited until recently [54, 55]. This is due not only to difficulties in the selection of effective solvents, but also to problems caused by the peculiarities of studies on concentrated solutions of cellulose. Very long periods of time may elapse before a system reaches equilibrium because of high viscosity.

Cellulose materials are known to be characterized by strong orientational van der Waals interactions and, often, by hydrogen bonds. The conformational calculations for isolated chains to determine the structural estimation of the rigidity of the cellulose skeleton are a necessary step when selecting solvents. Cellulose and its derivatives form lyotropic liquid-crystalline phases in non-aqueous solutions [56]. The most complete and detailed study has been carried out for solutions of hydroxypropylcellulose, acetates and nitrates of cellulose with high degrees of substitution [57–62].

A short review of the recent investigation of the phase equilibrium of lyotropic liquid crystals of polymers has been published by Suto [63]; in particular, he deals with liquid-crystalline polypeptides, toluene and dichloromethane solutions of poly(*n*-hexyl isocyanate), poly(γ-benzyl L-glutamate) and liquid-crystalline derivatives of cellulose.

The theory of phase separation in lyotropic liquid crystals and examples of experimental investigations of phase equilibria in solutions of the semi-flexible polymers based on poly(γ-benzyl L-glutamate) and dimethylformamide (DMF), on poly(*n*-hexyl isocyanate) and toluene or dichloromethane, or on hydroxypropylcellulose (HOPC) and dichloroacetate have been published by Sato and Teramoto [64].

The phase diagrams of solutions of cellulose and its derivatives were obtained using various methods and a combination of techniques: polarization optical microscopy (taking into account the temperature and concentration at which double refraction occurs), differential scanning calorimetry, rheology (judging the positions of the extremes in the concentration and temperature dependence of viscosity at low shear rate), NMR and PMR spectroscopy and light scattering. The results are unusual and differ noticeably from those obtained, for instance, for polypeptides. The variety of non-aqueous solvents for cellulose derivatives is wide, and the existence of an LC state in them has been proved beyond doubt [27, 59]. The question of the selection of solvents for realization of an LC state in non-aqueous solutions of cellulose is more complicated. On the one hand, it is limited by the difficulties of obtaining highly concentrated solutions, and on the other by the requirements of the destruction of fibrillar supramolecular structure of cellulose and of the formation of the mesophase incipients in the absence of macromolecular destruction by the non-aqueous solvent. It is no accident that few known cellulose systems satisfy these requirements. The LC state in cellulose–non-aqueous solvent systems is an exception rather than a rule.

A highly oriented state was not found in N_2O_4–polar aprotic solvent systems when cellulose is dissolved. The opinion of different authors [59, 63] on these solutions is the same: they are isotropic. There are contradictory opinions on the cellulose–lithium chloride–dimethylacetamide (DMAA) system and on cellulose solutions in *N*-oxides of ternary amides, in particular *N*-methylmorpholine-*N*-oxide. The phase diagram [59, 60] of the cellulose–LiCl–DMAA system (Figure 1.4) contains a region of LC state at cellulose concentrations above 10%.

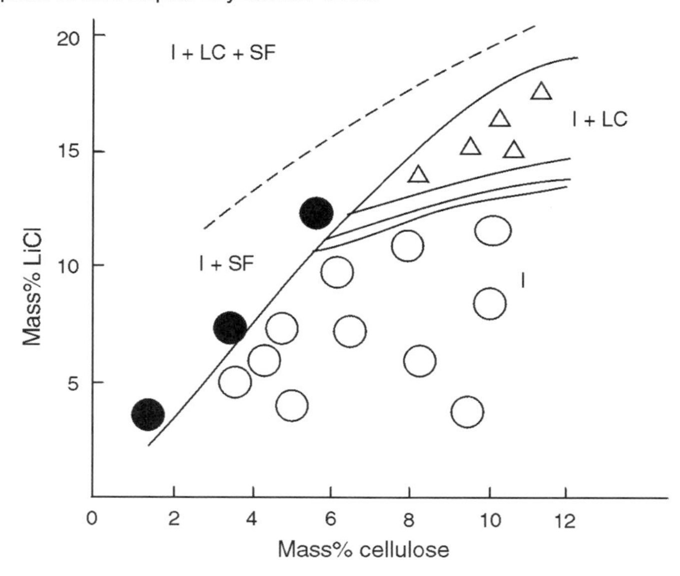

Figure 1.4 Phase diagram of the system cellulose–dimethylacetamide–lithium chloride. I, Isotropic solution; LC, liquid crystal; SF, solid fibre; I+LC+SF, three-phase region

There is also a more extensive region corresponding to isotropic properties and a narrow part of the diagram which is characterized by the entity having anisotropic and isotropic properties. However, there are criticisms of the results in the above-mentioned paper [55]. The phase diagram of poly(benzyl glutamate) solution in the same non-aqueous solvent is of similar form (Figure 1.5) [33]. The reviewers also considered factors which favour an additional increase in the rigidity of the cellulose macromolecule. Solid solutions have been experimentally found in the cellulose–DMAA–LiCl system and also in systems of aromatic polyamides in mineral acids [64]. These are not consistent with predicted phase diagrams based on available theoretical models.

Nishio and St John Manley [121] investigated blended films of cellulose and poly(vinyl acetate) (PVA) made from mixed solutions in *N,N*-dimethylacetamide–lithium chloride which were coagulated in a non-solvent. They used wide-angle X-ray diffraction (WAXD), differential scanning calorimetry (DSC) and dynamic mechanical measurements to characterize the state of miscibility of the clear films which were obtained over the whole composition range. The crystallinity of the PVA decreased with increasing proportion of cellulose. When the cellulose content was greater than 70 mass% there was no tendency for the PVA to crystallize. Good miscibility at cellulose contents greater than 60 mass% was interpreted as due to an increased interaction through hydrogen bonding of hydroxyl groups. This was

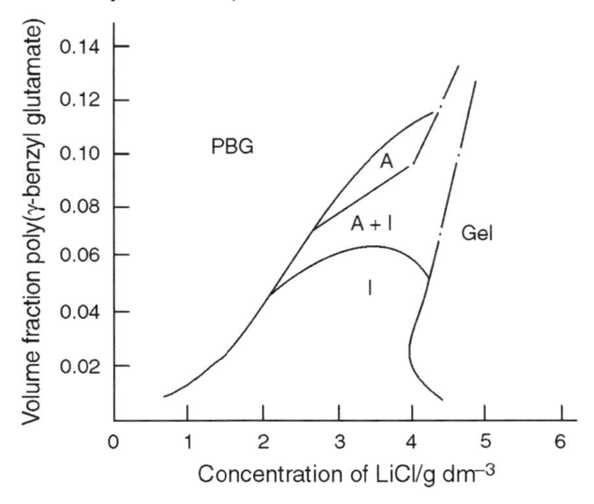

Figure 1.5 Phase diagram of system poly(benzylglutamate)–dimethylacetamide–lithium chloride

supported by evidence from DSC of decrease in melting and crystallization temperatures.

Nishio *et al.* [122] found a similar behaviour when blend films of cellulose and polyacrylonitrile were obtained in the same way using the same mixed solvent. In this case there was good miscibility when the cellulose content was greater than 50 mass%. This was interpreted as due to specific interaction between hydroxyl groups of cellulose and nitrile groups in the polyacrylonitrile.

Zhang and McCormick [123] prepared a series of derivatives of cellulose using *N,N*-dimethylacetamide as solvent. They successfully acylated cellulose with a series of unsaturated carboxylic acids or their anhydrides. Acylation with vinyl acetic acid, ethyl fumarate or cinnamic acid proceeded readily to form derivatives which were readily soluble in dimethyl sulfoxide (DMSO).

There is rheological evidence for mesomorphism in solutions of cellulose in methylmorpholine-*N*-oxide (MMO) [65,66] and for the appearance of a highly oriented state in the solutions of cellulose in MMO–DMSO mixture [61]. However there are grounds for believing that the formation of an LC phase is unlikely in these systems [59].

Petropavlovskii *et al.* [68] published convincing evidence for the destruction of the fibrillar structure of cellulose in amino oxides (e.g. *N,N,N*-trimethylamine-*N*-oxide) and of the appearance of a mesophase in the form of incipients.

In all cases considered, there is a discrepancy between the theoretical and experimental values of the critical concentrations (C^*) at which the phase with anisotropic properties appears. This is due to the use of incorrect values of the Kuhn segment in theoretical calculations.

Better agreement between the theoretical and experimental values is possible if the value of the statistical Kuhn segment is verified and if the influence of the molecular mass distribution in the non-aqueous solutions of cellulose is taken into account.

According to data given by Chanzy and Peguy [69], the critical concentration for the formation of anisotropic solutions of cellulose in MMO–H_2O depends on the molecular mass. The results of the phase analysis of the cellulose–MMO–H_2O system have been published [67]. Phase diagrams for non-aqueous solutions of cellulose in mixtures of CF_3COOH with CH_2Cl_2, $CHCl_3$ and $C_2H_4Cl_2$ at 298 K have been investigated [70]. All the systems studied possess some general regularities in the influence of cellulose concentration on the state of the solution. The nature and composition of the mixed non-aqueous solvents also show regular effects on solubility and state in solution.

Four regions can be distinguished in the phase diagrams:

- one-phase region of isotropic solution (I);
- two-phase region with anisotropic and isotropic properties (II):
- region of LC state (III);
- double-phase region: non-aqueous solvent–polymer (IV).

This is illustrated in Figure 1.6, which shows the phase diagram of the cellulose –CF_3COOH–$C_2H_4Cl_2$ system. Pure CF_3COOH has a rather limited solvency for cellulose of not more than 10 mass%. When chlorinated hydrocarbons are added, the solvency of the composite solvent for cellulose increases noticeably. The composition range in which mixtures of CF_3COOH and CH_2Cl_2, $CHCl_3$ or $C_2H_4Cl_2$ dissolve cellulose is limited to a mole fraction of the chlorinated hydrocarbon from 0 to an upper limit of between 0.5 and 0.7. The highest solubility is attained when the mole fraction of chlorinated hydrocarbon is between 0.2 and 0.4. For these mixtures the solubility of cellulose can be as much as 20 mass%.

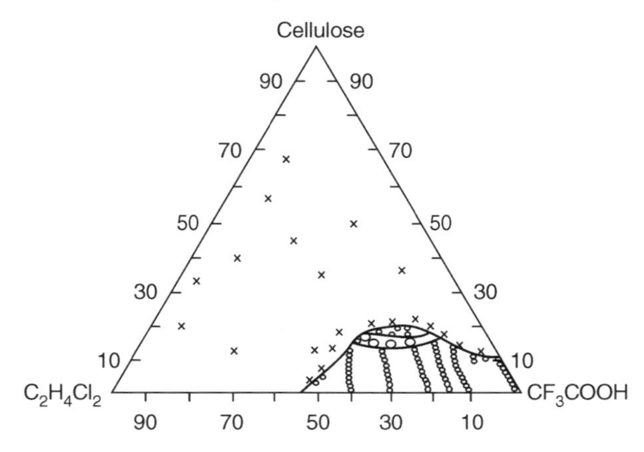

Figure 1.6 Phase diagram of system cellulose –CF_3COOH–1,2-$C_2H_4Cl_2$ at 298 K

Solubility decreases when the chlorinated hydrocarbon content is raised above 0.4 mole fraction. Cellulose is insoluble if $x_{CHCl_3} > 0.49$, or $x_{CH_2Cl_2} > 0.67$ or $x_{C_2H_4Cl_2} > 0.54$.

It is interesting to make further comparisons of the behaviour of cellulose in these mixed chlorinated hydrocarbon–CF_3COOH solvents.

At 298 K it is found that the region of the composition diagram in which LC order occurs widens in the sequence $CH_2Cl_2 < CHCl_3 < C_2H_4Cl_2$. The range of mole fraction of chlorinated hydrocarbon over which it extends is 0.15 for solvent containing CH_2Cl_2, 0.20 for $CHCl_3$ and 0.32 for $C_2H_4Cl_2$. The critical concentration of cellulose in the following mixtures, of approximately similar composition, $CF_3COOH–C_2H_4Cl_2$ ($x_{C_2H_4Cl_2} = 0.32$), $CF_3COOH–CHCl_3$ ($x_{CHCl_3} = 0.29$), CF_3-$COOH–CH_2Cl_2$ ($x_{CH_2Cl_2} = 0.36$), is 18–20 mass%. This corresponds to a mole fraction of polymer in the non-aqueous solutions of $1.4 \times 10^{-4}–1.5 \times 10^{-4}$. In addition to increasing the solubility of cellulose in CF_3COOH, the addition of chlorinated hydrocarbons has a further effect, i.e. it decreases the initial Newtonian viscosity of the solutions.

The spontaneous occurrence of double refraction [56, 62] and manifestation of the phase transitions in the systems studied in the state of rest depend on the interparticle interactions in solutions. The increase in the polymer chain rigidity due to partial trifluoroacetylation of cellulose or to the formation of an H-complex of cellulose with the mixed solvent can become a crucial factor favouring a highly oriented state of the system. Self-association of CF_3COOH in mixtures with chlorinated hydrocarbons has a considerable influence on the strengthening of donor–acceptor interaction of polymer with the non-aqueous solvents and on the character of molecular mass distribution.

A stable lyotropic mesophase was found in 6% solutions of cellulose in a $CF_3COOH–CH_2Cl_2$ mixture (60:40 by volume), and in 8% solutions in a 70:30 mixture [71]. The latter mixture is the best solvent, causes minimal destruction of the cellulose and solutions have the lowest viscosity [70,71]. Solutions of cellulose of 12 and 14 mass% have a nematic structure in the shear field.

Nishio and St John Manley [116] described the preparation of blends of cellulose with nylon 6 and with poly(ε-caprolactone) (PCL) from solution in *N,N*-dimethylacetamide–lithium chloride by coagulation in a non-solvent. In the case of the blends with PCL they found evidence of an amorphous phase in which a homogeneous mixture was formed. There was also a significant shift to higher temperatures of the mid-point of the PCL glass transition region when the concentration of cellulose was 20 and 30 mass%. The phenomena were interpreted in terms of an optimum density of interacting hydroxyl and ester groups for true miscibility of cellulose.

Nishso *et al.* [117] reported similar phenomena in the case of cellulose–poly(vinyl alcohol) blends which indicated that hydrogen bonding could lead to a favourable interaction between two different polymers leading to greater miscibility.

Masson and St John Manley [118] prepared blends of cellulose with polyvinylpyrrolidone and reported that X-ray analysis showed that each polymer influenced the structural order of the other.

Publications giving phase diagrams of non-aqueous solutions of cellulose ethers and esters are more numerous. Various forms of phase diagrams are possible for cellulose derivatives depending on the type of the substituent that is introduced into the macromolecules and on the nature of the solvent.

The phase diagram of system cellulose triacetate (CTA)–acetic acid–methylene chloride has been published by Iovleva [72], who distinguished between two areas in the Gibbs triangle: one of homogeneous solution and the other of heterogeneous mixture. The latter consists of a region where there is a clear division between the phases and the intermediate states which form a gel. These intermediate states consist of systems with incomplete division between phases. The boundary between solution and the region where there is a clear division between phases is well defined, especially at low polymer concentration. The boundary between solution and gel is less well defined. The region corresponding to solution becomes greater with increase of temperature. The influence of molecular mass was not considered in detail, it being stated only that the region of the solutions becomes smaller with increasing molecular mass.

The phase diagram of CTA–methylene chloride–ethanol system with analogous phase transitions has been published by Malafeyeva and Averyanova [73].

Studies of concentrated and dilute solutions of CTA in a mixture of methylene chloride and methanol with addition of saccharose octabiphenylphosphate (SOBPP) have been reported by Derzhavina and Zakurdayeva [74]. An increase in SOBPP content in the solutions resulted in increasing statistical rigidity of CTA macromolecules and in the appearance of a liquid-crystalline structure.

Phase diagrams of cellulose acetate–acetic acid–water and cellulose acetate–acetone–water systems have been published [75, 76]. These diagrams also have small regions in which the state of the system is homogeneous. These depend on the nature of the solvent and increase on going from acetone solutions to acetic acid solutions. The difficulties in differentiating disperse phases from metastable solutions of acetylcellulose in the diethyl ether of diethylene glycol have been considered by Abaturova *et al.* [76]. In this case the diagram has well defined areas in which a single-phase stable solution exists.

Werbowij and Gray [77] were among the first to show that cellulose derivatives could form liquid crystals and publish the corresponding phase diagrams. At about the same time, Luise *et al.* [78] reported that the cellulose ethers and esters in organic solvents formed anisotropic solutions at 298 K. Manifestations of anisotropy for the derivatives with a degree of substitution $\gtrsim 1.7$ were observed at concentrations $\gtrsim 20$ vol.%. These are examples of a more general phenomenon. Recent investigations have shown that macromolecules of cellulose derivatives can possess an increased equilibrium rigidity compared with that of the cellulose macromolecules [79].

For cellulose the value of the Kuhn segment is 100–140 Å, for CTA it is 260 Å and for trinitrate 330 Å. This is evidently related to the decrease in, or possibly complete absence of, intra- and intermolecular hydrogen bonding in the cellulose derivatives. This makes it possible, in many cases, for the derivatives to be dissolved in solvents which are comparatively weakly active in the breaking of hydrogen bonds.

A complete phase diagram of the CTA–trifluoroacetic acid–water system was obtained by Meeten and Navard [80]. In this case, there occur a multiple phase transition and five types of regions, including isotropic solution, gel, double-refracting gel, turbid gel and CTA–solvent double-phase region. The absence of the double-phase region with isotropic and anisotropic properties is a peculiar feature of this diagram. In mixtures containing CF_3COOH the authors observed isotropic and clear solutions at a water content below 30%. At a water content above 70% the mixed solvent did not dissolve CTA.

The formation of a mesophase in CTA–CF_3COOH–chlorinated hydrocarbons systems has been reported [81]. Figure 1.7 shows the experimental phase diagram of CTA in CF_3COOH–$CHCl_3$ mixtures at 298 K [62]. As can be seen, the CTA solubility increases considerably, from 8 to 35 mass%,when trifluoroacetic acid is added to chloroform. The solvency is highest when the mole fraction of chloroform is between 0 and 0.49. In this range of solvent composition multiple phase

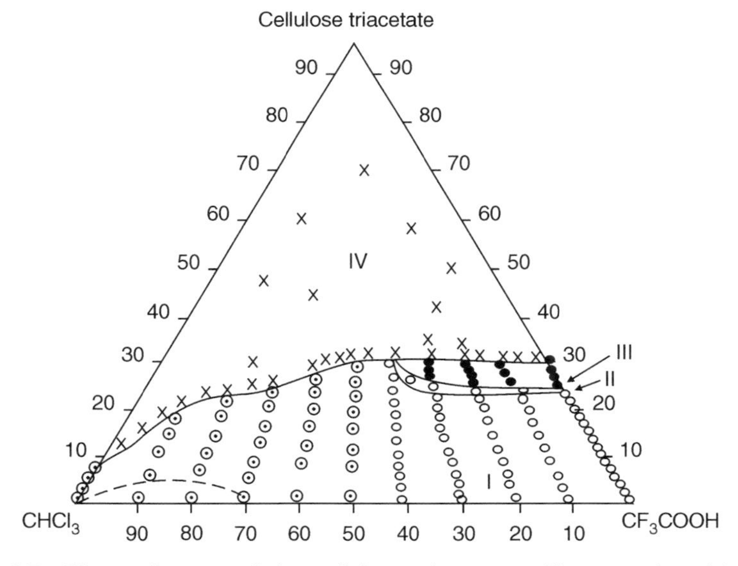

Figure 1.7 Phase diagram of the cellulose triacetate–trifluoroacetic acid–chloroform system at 298 K. I, Single-phase region of isotropic solution; II, double-phase region with isotropic and anisotropic properties; III, region of LC state; IV, double-phase region, non-aqueous solvent–fibre

transitions occur with increasing polymer concentration. In the phase diagram one can distinguish between the region in which the solution has isotropic properties (I), a wide region corresponding to the presence of gels (II) and a limited two-phase region where there is co-existence of the phase with isotropic properties, a narrow region of LC state (III) and (IV) a two-phase region of solid state (fibres) and non-aqueous solvent. The LC phase was found by use of a polarization microscope. It was characterized by double refraction and possesses the optical properties of a cholesteric liquid crystal. In the crossed Polaroids of the microscope, it was possible to identify certain types of ordering corresponding to interchanging bands which changed their form when the polarization angle was changed by 90°. The LC phase is stable over the temperature interval from freezing to melting point of the liquid crystals. At the melting point there occurs a phase transition to a liquid state with isotropic properties.

It is interesting to compare phase diagrams of non-aqueous solutions of cellulose with those of cellulose triacetate. Unlike cellulose, CTA forms solutions over the entire range of composition of the binary trifluoroacetic acid–chlorinated hydrocarbon solvent. It is also soluble in the pure chlorinated hydrocarbon and in the pure triacetic acid. Unlike cellulose, CTA can exist in an LC state in pure trifluoroacetic acid. The tendency of CTA to exist in an lyotropic LC state in various mixtures increases in the order $CF_3COOH–CHCl_3 < CF_3COOH–C_2H_4Cl_2 < CF_3COOH–CH_2Cl_2 < CF_3COOH$ and for cellulose $CF_3COOH–CH_2Cl_2 < CF_3COOH–CHCl_3 < CF_3COOH–C_2H_4Cl_2$.

Refractive indices of solutions have been measured to verify the concentration–temperature boundaries of phase transitions. A change in the variation of the refractive index with change in concentration was found. This is related to the fluctuation of orientations at critical concentration of polymer [70].

X-ray diffraction studies at low angles reveal the existence of intensive Bragg reflections corresponding to 9–11° 2θ.

Myasoedova and co-workers have made X-ray measurements at low angles on concentrated solutions of CTA in trifluoroacetic acid and mixtures containing chloroform (mole fraction 0.29) or dichloromethane (mole fraction 0.365). These studies provided information about the orientational ordering characteristic of LC systems. X-ray measurements on films of polymer, regenerated from the solutions, were also carried out.

The low-angle reflections probably correspond to periodic repetitions of the structural elements of CTA molecules and indicate the existence of near monomeric order in CTA solutions [62]. Intensive Bragg reflexes in the scattering curves that are characteristic of the films which are regenerated from 15% solutions of CTA, in the 9–11° 2θ region indicate that the mesomorphous structure in preserved in the solid films.

Chromatograms obtained by gel permeation chromatography of solutions in acetone display three peaks [119]. The first peak to be eluted corresponds to a microgel of cellulose acetate molecules. The second peak corresponds to dispersed

Table 1.1 Experimental data from X-ray diffraction analysis of non-aqueous solutions of CTA and of films regenerated from these solutions

System	$x_{CTA} \times 10^4$	$\bar{a}$/Å	L/Å	$A/\bar{a}$	γ_m/Å
$CTA-CF_3COOH$	1.9	8.63	23.62	0.19	37.12
$CTA-CF_3COOH-CHCl_3$	2.6	8.43	25.33	0.18	39.76
$CTA-CF_3COOH-CH_2Cl_2$	2.5	8.85	25.32	0.19	39.76
$FilmCTA-CF_3COOH-CHCl_3$	1.0	8.85	20.90	0.21	32.83
$FilmCTA-CF_3COOH-CH_2Cl_2$	0.9	8.85	20.90	0.21	32.83

molecules containing anionic groups which are possibly sulfuric acid groups. The third and main peak corresponds to the main bulk of the polymer.

Tang *et al.* [120] investigated samples of cellulose acetate prepared by solution acetylation. Scanning electron microscopy and transmission electron microscopy showed that samples prepared at different temperatures had different morphologies. This indicated that there were different mechanisms of acetylation at low (45 °C) and high (90 °C) temperatures.

The X-ray diffraction data were used to calculate the dimensions of separate ordered regions (crystallites), L, interaction radii, γ_m, and values of the disordering parameters, $A/\bar{a}$, for all the systems studied. The results of the measurements are given in Table 1.1. The closeness of values of $\bar{a}$ (average distance between the neighbouring chains) proves the identity of orientational structures in solution and in film.

Circular dichroism (CD) spectra using an external chromophore were used to identify the type of mesophase formed by CTA in a $CF_3COOH-CHCl_3$ mixture and other non-aqueous solutions. The dye 1-ω-diethylaminopropylamido-4-hydroxy-9,10-anthracenedione hydrochloride was introduced as the external chromophore.

The dye was dissolved in the mesophase formed and the absorption and CD spectra of the solution obtained were measured. Figure 1.8 shows the CD spectra that were measured in the region of absorption of the dye. It can be seen from Figure 1.8, that an intensive positive band appears in the region of dye absorption ($\lambda = 475$ nm). The high intensity of the band points to the appearance of anomalous optical activity of the external chromophore in the LC phase at $C_{CTA} > C^*$.

The appearance of such a band cannot be explained by the optical properties of isolated dye molecules. Not only is the formation of an LC phase necessary for such a band to appear but also the presence of long-range spiral order of the dye molecules is needed. This ordering of the dye molecules is caused by spiral packing of pseudonematic layers that are formed by CTA molecules. Substitution of chloroform by methylene chloride leads to the change in the direction of spiral twisting.

Navard and Haudin [82] and also Bheda *et al.* [83] reported the formation of the lyotropic LC state by CTA solutions in dichloroacetic acid and nitromethane. It has

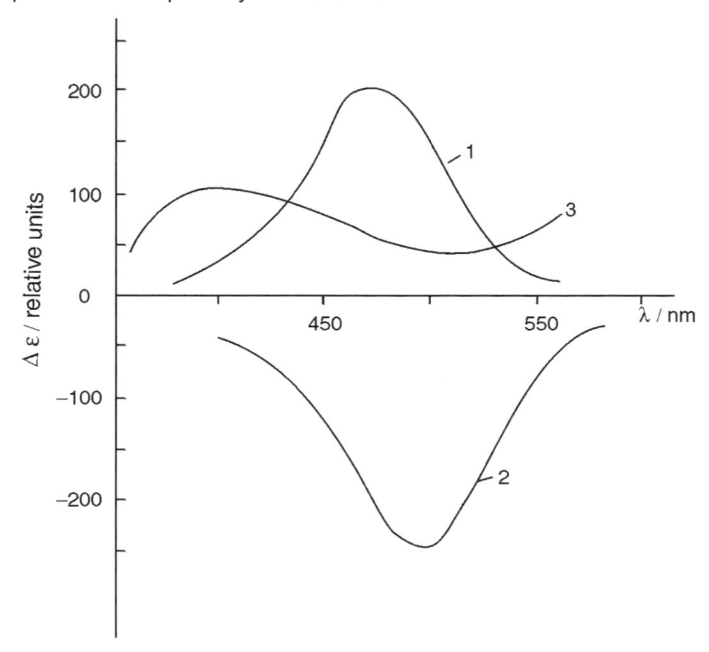

Figure 1.8 Circular dichroism spectra of 30% solutions of CTA in mixed solvents. (1) $CF_3COOH-CHCl_3$ ($x_{CHCl_3} = 0.24$); (2) $CF_3COOH-CH_2Cl_2$ ($x_{CH_2Cl_2} = 0.36$) in the presence of an 'external' chromophore; (3) 30% solution of HOPC in *N*-methylpyrrolidone

also been reported by Bianchi *et al.* [84] that solutions of cellulose diacetate with a degree of substitution of 2.3 in dimethylacetamide can form a mesophase. The authors showed that here the transition to the liquid-crystalline phase proceeds through a narrow region consisting of a mixture of isotropic and anisotropic phases.

Molecular ordering in liquid-crystalline solutions of cellulose triacetate has been studied by the use of NMR spectroscopy. It has been shown that LC solutions in the system $CTA-CF_3COOH-C_2H_4Cl_2$ have a very small step in the spiral. Even a strong magnetic field is insufficient for the transition of the cholesteric CTA structure into the nematic structure. The NMR spectrum of CTA in $CF_3COOH-C_2H_4Cl_2$ (70 : 30 by volume) shows evidence for the nematic nature and indefiniteness of the spiral step [85].

Results of studies on the phase transition on heating of liquid-crystalline solutions of cellulose diacetate (CDA) and CTA in CF_3COOH are of interest [86, 87]. The dependence on polymer concentration and the temperature of the region in which the anisotropic phase exists was determined. There is a certain temperature hysteresis, since the transition into the isotropic state on heating and the reverse transition on cooling occur over a temperature interval. Similar measurements have been made on solutions of cyanoethylcellulose in CF_3COOH and dimethylformamide [88, 89].

Kamide *et al.* [36] attempted to formulate a general theory of the mechanism of the influence of the nature of the solvent on the occurrence of highly oriented states in solutions of cellulose nitrate, CTA and cellulose acetate (CA). The values of the interaction parameter of Flory and Huggins, $\chi_{1,2}$, were used as a criterion because this parameter reflects the thermodynamic quality of the solvent. The influence of the nature of the solvent on the value of the critical concentration, C^*, is usually evident. The formation of lyotropic liquid crystals of cellulose acetates was studied in more than 30 solvents. The critical concentration was shown to be linearly dependent on pK for acidic solvents. The influence of the molar volume of the solvent on C^* values was considered in other systems.

The influence of the degree of substitution (DS) and of degree of polymerization (DP) and of temperature on C^* has also been investigated by Dayan *et al.* [90]. For example, C^* for cellulose acetate in dioxane has been found to vary between 43 and 55% over the temperature range 25–62 °C. The authors showed that a decrease in DS lowered the transition temperature and also the concentration at which transition occurred. Similar results showing the influence of DP on C^* have been published by Navard and Haudin [65].

For aromatic *P*-polyamides

$$V^* \sim 1/M$$

where V^* is the critical volume fraction for phase transition. The dependence of V^* on M for the solutions of the cellulose derivatives is less marked and agrees qualitatively with data given by Khokhlov and Semenov [91]. This shows that cellulose derivatives have lower rigidity than aromatic polyamides.

Reports of the lyotropic behaviour of cellulose nitrates in a variety of individual and mixed solvents have appeared [92]. Phase diagrams of rigid rods–solvent–non-solvent multicomponent systems testify to the division into isotropic and anisotropic phases in agreement with Flory's theory. The phase diagram can contain at least 10 regions (Figure 1.9) including those with the signs of co-existence of two or more phases (among them liquid crystals and crystallosolvates). The textures demonstrated by the anisotropic phase of, for example, the cellulose (DS = 2.72) nitrate–tetrahydrofuran–ethanol system are typical of Schlieren-type nematics divided by Grangjean partitions that keep the cholesteric order.

It is interesting that the textures observed in crossed Polaroids are characteristic of a parabolic focal–conic type and are similar to those of lyotropic hydroxypropylcellulose. It is noteworthy that both cellulose and its derivatives, being chiral, tend to have cholesteric supramolecular ordering in solution. In the case of the systems discussed above, the values of the critical concentration for mesophase formation indicate that the polymer molecules are not ideally rigid. The formation of laminated structures is possible in these systems. This is determined

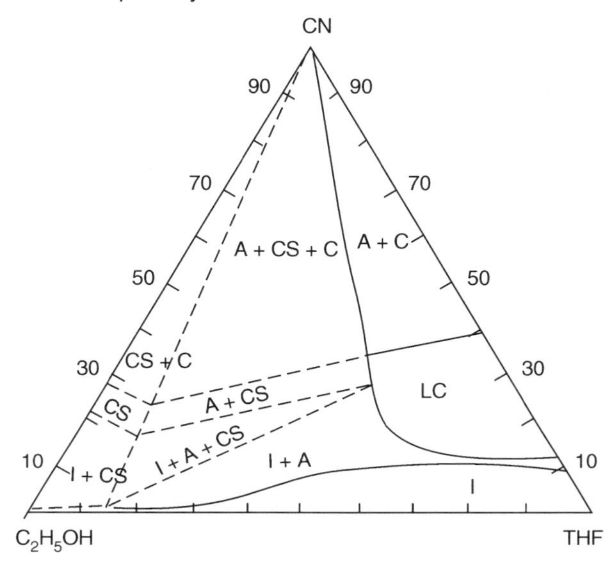

Figure 1.9 Phase diagram of the cellulose nitrate–ethanol–tetrahydrofuran system at 298 K [92]. A = anisotropic; C = crystalline; CS = crystallosolvate; I = isotropic; LC = liquid crystal

by the chiral nature of the molecules. Materials regenerated from solution do not contain quantitatively definable traces of crystallosolvate. Non-periodic layers of crystallites can occur in the condensed state owing to suturing between the chains of cellulose nitrate.

Values of the critical concentration, C^*, calculated by Gilbert [93], are 0.7 for cellulose and 0.27 for HOPC. Other derivatives have C^* values lying between these values. He also found peculiarities in the cholesteric structure and evidence of a forced transition into the nematic structure under the influence of shear deformation or of an electric field. He stressed that cellulose derivatives have a tendency towards lyotropic mesomorphism.

Werbowij and Gray [94] found that 50–80% solutions of HOPC in water exhibit double refraction. This occurred at concentrations considerably exceeding those predicted from Flory's theory. In a thin layer, such solutions possess complex indications of the LC state including optical activity characteristic of Grangjean texture of a cholesteric mesophase. The step of the over-molecular spiral was calculated to be 72 Å.

Textures occurring at shear of a thin (15 μm) layer of HOPC solutions in water and methanol were investigated by Donald *et al.* [95]. Bands oriented perpendicular to the direction of shear were formed in LC solutions when they were deformed at a shear rate $\gamma \approx 20\,\text{s}^{-1}$. The distance between them increased with increasing concentration and molecular mass of HOPC. A series of microphotographs, taken during systematic change in the polarization angle, were published. Judging by its appearance, the texture was identified as parabolic confocal.

It pointed out by Werbowij and Gray [96] that HOPC is amphiphilic and is a surface-active substance in aqueous solutions. There is therefore always the possibility of the formation of liquid crystals similar in type to biomembranes, soaps and block copolymers, in which the occurrence of a mesophase is not related to the rigidity of macromolecules.

Werbowij and Gray [96] and later Conio *et al.* [97, 98] obtained phase diagrams for solutions of HOPC and CA in dimethylacetamide. The phase diagram obtained by Werbowij and Gray [96] showed that it was possible to observe a double-phase region with isotropic and anisotropic properties when concentrations were between 39 and 47 mass%.

Myasoedova and co-workers have experimentally obtained the phase diagram of the HOPC–*N*-methylpyrrolidone system (Figure 1.10). In the region labelled **I** solutions are isotropic and in the region labelled **A** they possess anisotropic properties. If the phase diagrams of the HOPC–water, HOPC–ethanol, HOPC–DMAA and HOPC–*N*-methylpyrrolidone systems are compared, the influence of

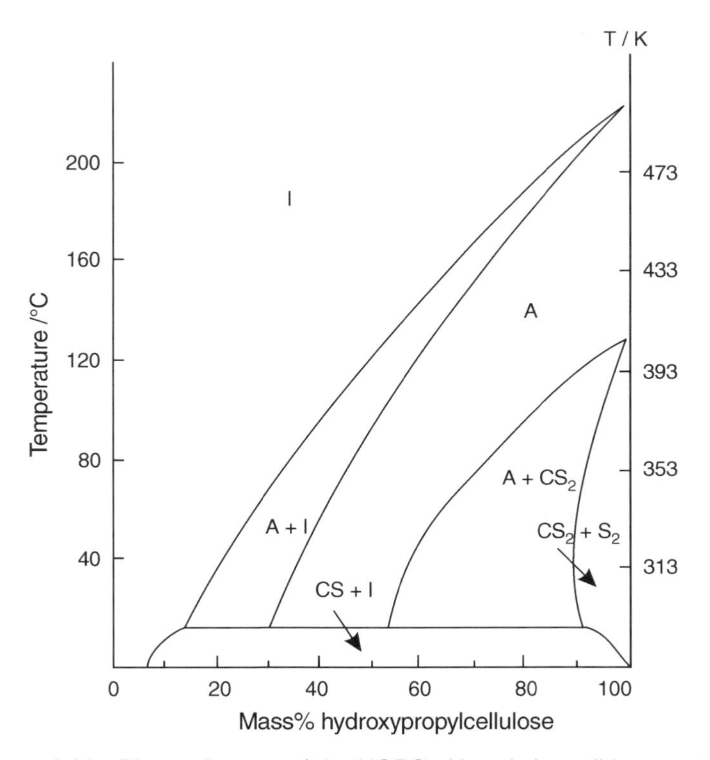

Figure 1.10 Phase diagram of the HOPC–*N*-methylpyrrolidone system

the nature of the solvent on the phase equilibria in polymer solutions is apparent. The results obtained show that at temperatures above 40 °C, the polymer precipitates from HOPC–water. There is no temperature limit to the solubility for solutions of HOPC in DMAA over the concentration range 50–80% which was investigated. It has been reported [99] that HOPC–DMAA solutions are coloured and exhibit double refraction at higher concentrations of HOPC than in aqueous solutions. This confirms that the solvent influences the spiral step. The lyotropic LC state of HOPC in acetic acid, water and mixed solvents has been investigated by Nishio *et al.* [100].

Davè *et al.* [124] investigated a system consisting of the flexible polymer hyaluronic acid (HA), the semi-rigid polymer (hydroxypropyl)cellulose (HPC) and water. The phase diagram of the ternary mixtures showed a homogeneous region when the overall polymer concentration was lower than $C_p \approx 15\%$. At higher concentrations two demixing areas were found. The first, in the range $C_p = 15-21\%$, corresponds to an equilibrium between two isotropic phases. When $C_p > 21\%$ a demixing area corresponds to a HPC mesophase co-existing with an isotropic phase of both polymers.

Phase equilibria in the ethylcellulose–acrylic acid system was investigated by Zugenmaier [101]. Figure 1.11 shows part of the phase diagram in the form of a corridor with three distinct regions: isotropic solution, mixture of phases with isotropic and anisotropic properties and an anisotropic solution. This diagram is

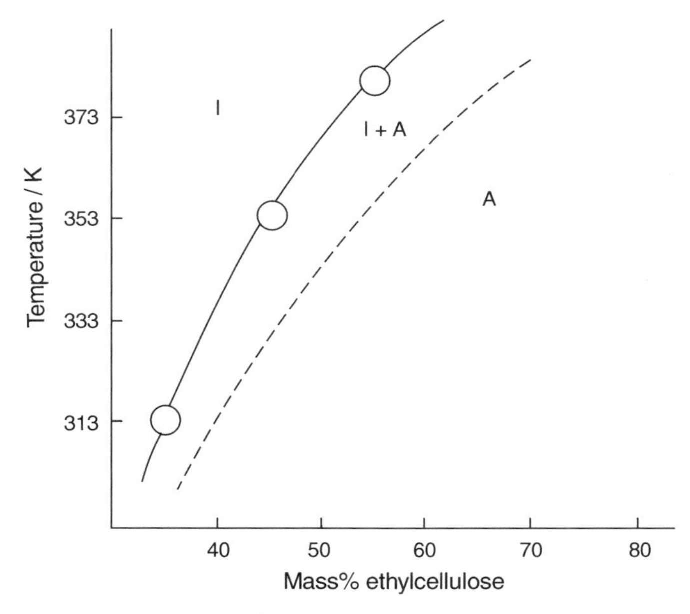

Figure 1.11 Phase diagram of the ethylcellulose–acrylic acid system

characterized by a widening region of the two-phase state when the temperature increases and by gel formation at high temperatures. The behaviour indicates that the mesophase which is formed is cholesteric.

The structural characteristics of cellulose tricarbanilate (CTC) and ethylcellulose (EC) in crystalline and liquid-crystalline states were investigated by Huang *et al.* [102]. Lyotropic systems were formed by CTC with butanone or 2-pentanone as solvents and by EC in glacial acetic acid. In the anisotropic solutions, positive double refraction was detected for CTC in both solvents and negative double refraction for EC in glacial acetic acid. In the case of CTC in both solvents, the spiral step increases with increasing temperature and concentration, the twisting of the cholesteric spiral being right-handed. EC solutions are characterized by the left-hand twisting and by the spiral step decreasing with the increase in temperature and concentration. The typical value of the spiral step for the ethylcellulose–acetic acid system is 500 nm and for the CTC–butanone system 2000 nm. It was found possible to obtain any intermediate magnitude of the step by using mixed solvents.

Ethylcyanoethylcellulose (ECEC) is soluble in many solvents and forms liquid-crystalline solutions above the critical concentration. As the concentration increases, the solution changes from an isotropic into an anisotropic state via a two-phase system. The transition to a single phase system was observed in two-phase solutions of ECEC in dichloroacetic and trifluoroacetic acids. In these solutions, the LC structure in some cases changes into a spherulitic structure. For most organic solvents, the critical concentration C^* decreases with decreasing parameter of polymer–solvent interaction. In the case of organic acids, C^* decreases with increasing acidity.

Cellulose 3-chlorophenylurethane forms a liquid-crystalline phase in solutions of diethylene glycol monoethyl ether, methyl ethyl ketone, 2-pentanol, etc. [103]. Anisotropic solutions show positive double refraction and a cholesteric structure with a 360–380 nm step at a concentration of 0.7 g ml^{-1} at 25 °C. The magnitude of the step depends on the solvent nature and concentration. The value of the parameter of order for the layer of the highly oriented solution on the polymer matrix is 0.7 at sufficient remoteness from the limpidity point.

Investigations [104] devoted to the studies on double refraction in concentrated solutions of methyl, ethyl, cyanethyl and hydroxypropylcellulose in aqueous solutions of hydrochloric, nitric, phosphoric and perphosphoric acids are interesting. It is apparent that the type of phase diagram for cellulose and its derivatives depends on the nature of the substituent that is introduced into a macromolecule, on DP and DS and on the nature and composition of the non-aqueous solvent.

Investigations of phase equilibria and texture of the ternary system hydroxypropylcellulose–ethylcellulose–acetic acid (Figure 1.12) have been carried out by Laivins and Sixou [105].

The formation of ordered phases of HOPC in compatible and incompatible isobutyric acid–water mixtures has been studied by Gray [106]. Gas–liquid

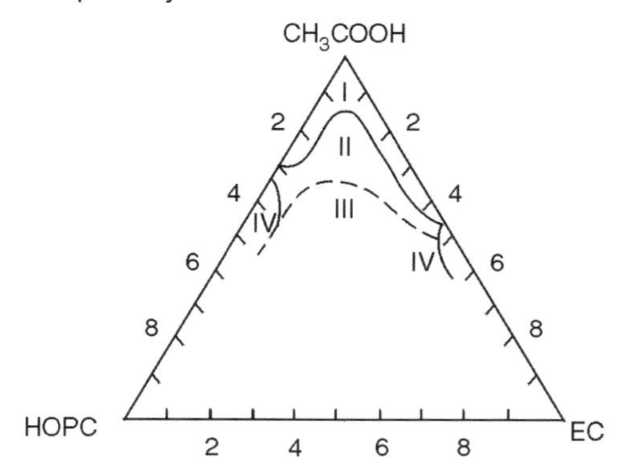

Figure 1.12 Phase diagram of the HOPC–EC–CH_3COOH system. I, Isotropic monophase; II, isotropic–isotropic biphase; III, isotropic–anisotropic and anisotropic–anisotropic biphase; IV, anisotropic monophase

chromatography and polarization microscopy were used for the analysis. The phase diagram of HOPC–isobutyric acid mixtures exhibits six types of homo- and heterogeneous states of the system. The dependence of transitions in the system on HOPC concentration, mixed solvent composition and temperature were discussed. Optical and rheological properties of the ternary systems were compared with those of homogeneous solutions in acid and water.

Various aspects of investigation and application of the liquid-crystalline solutions and melts of cellulose and its derivatives have been generalized by Shimamura *et al.* [107]. The structure of the elementary links of cellulose and the influence of the nature of substituents replacing OH groups on the rigidity of macromolecules is considered. On the one hand, the decrease in the fraction of OH groups facilitates structural rearrangements due to the decrease in the ability to form H-bonds. On the other hand, for trisubstituted celluloses there is an increasing tendency for crystallization and gelation to occur. This is exhibited the most markedly in certain solvents.

A better understanding of the conditions necessary for the formation of anisotropic phases in lyotropic systems based on cellulose or its derivatives will eventually have important practical applications. Prediction of conditions for obtaining highly orientated hard-wearing fibres and films from liquid crystalline solutions and melts will soon be possible.

2 THERMOTROPIC MESOMORPHISM OF CELLULOSE DERIVATIVES

The first evidence for the existence of a cellulose derivative in a thermotropic cholesteric mesophase was found in HOPC melts by Jiang [108].

HOPC with a molecular mass of 6×10^4 is a liquid crystal between 433 and 478 K. On heating, the DSC curve exhibits one peak and on cooling two peaks [109]. This is characteristic of monotropic liquid crystals. Enthalpies of crystalline–isotropic transition and of transition from solid into liquid-crystalline state were calculated, assuming that HOPC is a mixture of crystalline and LC phases at room temperature.

Myasoedova and co-workers have experimentally studied the influence of molecular mass and of the degree of substitution of HOPC on the temperatures and enthalpies of phase transitions. Figure 1.13 shows the dependence of the enthalpies of isotropic–anisotropic transitions of HOPC on the molecular mass. The dependence was found by DSC. As can be seen from Figure 1.13, an increase in molecular mass from 6×10^4 to 100×10^4 results in a decrease in the enthalpy of anisotropic–isotropic transition by an order of magnitude. Samples with a degree of molar substitution of three (Figure 1.14) are characterized by the widest temperature range in which the LC phase exists. A further increase in the degree of substitution leads to the narrowing of the range.

Solid HOPC samples with molecular mass of 6×10^4 were prepared by three different methods. One method involved the formation of films from LC solutions in water and the second the formation from an anisotropic melt. The third method involved dissolution in 2-hydroxyethyl methacrylate with subsequent polymerization of the solvent [109]. Samples of solid ethylcellulose were prepared by a similar

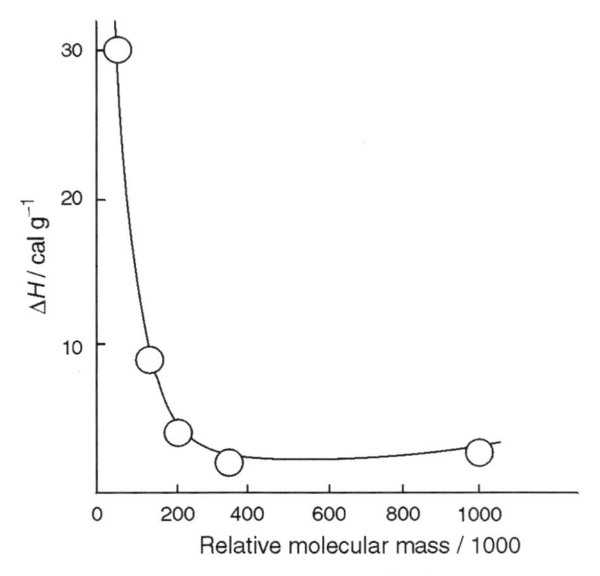

Figure 1.13 Variation with molecular mass of HOPC of the enthalpy of isotropic–anisotropic transition

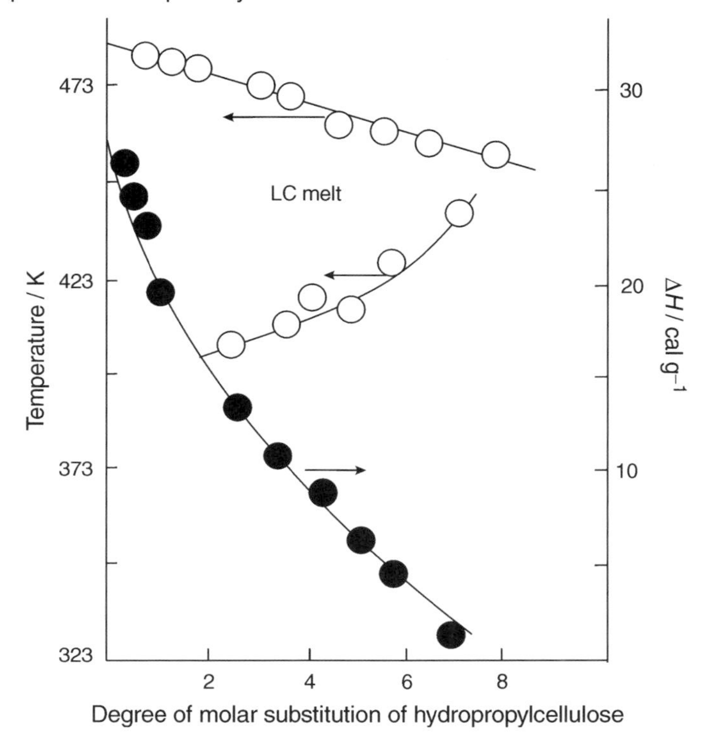

Figure 1.14 Variation with degree of molar substitution of HOPC of the enthalpy and temperature of isotropic–anisotropic transition

method to this last method. Ethylcellulose was dissolved in acrylic acid which was subsequently polymerized to give a composite ethylcellulose–polyacrylic acid.

Samples formed under conditions of different extents of shear deformation were investigated by electron microscopy. Round particles of 2000 Å diameter, which form rouleaus and disk-like lamellae, are a typical element of the cellulose systems. One can follow the transformation of the initial structures while the extent of deformation is increasing. Lengthening and bending of the rouleaus and also fibrillation may be observed. It is possible to follow the appearance of the single-axis order in fibrils and subsequent change into variants with a flat zig-zag or folded shape.

The conditions for the formation of the liquid-crystalline phase in melts of HOPC and its derivatives have been studied by Shibayev and Yekayeva [110]. A series of *n*-aliphatic derivatives of HOPC were obtained via synthesis of its ethers. Optical investigations have proved that the first members of the homologous series form a thermotropic LC phase of cholesteric type. The dependences of the step in the cholesteric spiral on the number of carbon atoms in the aliphatic substituents, on the molecular mass of the polymers and on temperature were studied. The authors also reported studies of structures by X-ray diffraction measurements.

Acetoacetoxypropylcellulose (AAPC) with a degree of substitution 2.20 was obtained by the action of diketene–acetone adduct on HOPC dissolved in *N*-methylpyrrolidone [111]. The sample was investigated by polarization microscopy, DSC, IR and UV spectroscopy, refractometry, circular dichroism measurements and X-ray diffraction. Thermotropic transitions in AAPC were found at 400 K (crystal–nematic transition) and 447 K (limpidity point) with enthalpy changes of -13.2 and $-1.4\,\mathrm{J\,g^{-1}}$, respectively. On cooling, the transition temperatures decreased to 420 and 355 K. In a thin layer, anisotropic melts of AAPC exhibit predominantly a yellow–green colour in polarized light; this testifies to a cholesteric structure with a small spiral step (273 nm).

Acetoxypropylcellulose in the form of a thermotropic cholesteric polymer can also be synthesized from hydroxypropylcellulose and acetic anhydride [112].

Samples of hydroxypropylcellulose trifluoroacetate are cholesteric between room temperature and 383 K. The existence of the cholesteric structure at 298 K is especially interesting. Although the polymer possesses viscous-flow properties at this temperature, the textures are very stable for several months [113]. Optical microscopy reveals the appearance of characteristic striped textures in samples of thermotropic hydroxypropylcellulose trifluoroacetate between flat-parallel limpid plates. Stripes perpendicular to the direction of shear were formed either at high shear rate ($\gamma \approx 20\,\mathrm{s^{-1}}$) or after stopping the flow at low rates. Superposition of deformation on a sample which has developed a striped texture results in the destruction of the stripes.

It has been shown that, in addition to the transition layers, within each stripe there is orientation of long axes of macromolecules at an angle of $+45°$ to the direction of shear. The relaxation properties of long chains give rise to elastic effects and are considered to be the cause of such behaviour. It is found that the process of formation of domainic texture is not directly connected with the cholesteric nature of a cellulose derivative. It can, however, be initiated by longer relaxation processes that are characteristic of the forced transition of the nematic into the cholesteric structure.

Much attention has been paid to the problem of phase separation in polymer blends. Texture and morphology are important. Navard and Zacharades [113] published a short review of the basic experimental methods for characterizing cholesteric mesophases and phase transitions. This is followed by a discussion of the behaviour of mixtures of liquid-crystalline derivatives of cellulose with other polymers.

Pawlowski *et al*. [114] synthesized three cellulose esters: phenylacetoxy- (PAC), β-4-methoxyphenylacetoxy- and *n*-tolylacetoxycellulose, and two cellulose ethers, trimethylsilyl- (TMSC) and *tert*-butyltrimethylsilylcellulose (BTMSC). The aim was to create easily regenerated systems in order to obtain cellulose hydrate fibres and films. Polarization–optical and thermodynamic investigations showed that these cellulose derivatives are thermotropic liquid-crystalline polymers. However, the temperature interval for the existence of the mesophase is rather narrow. For PAC, the crystal–mesophase transition occurs at 388 K and the mesophase–isotropic solution transition at 398 K. BTMSC does not turn into the anisotropic state either in the melt or in solution. Other polymers form lyotropic liquid-crystalline systems at concentrations in solution of 44, 48, 50 and 27 mass%, respectively. The presence in the texture of 'fingerprints' together with dispersion of optical rotation indicated that the structure of the mesophases formed was cholesteric. Regeneration of the mesophase derivatives was not difficult. It was possible to orientate an anisotropic system and obtain the material from cellulose hydrate.

Kamide *et al.* [103] synthesized cellulose phenylacetate and 3-phenylpropionate and their ability to form a thermotropic LC phase was investigated. Both polymers turn into the LC state with limpidity temperature 440–450 K. Fibres formed from the mesophase melts of these cellulose derivatives have a striped appearance in polarized light. The stripes are oriented perpendicular to the direction in which the fibres are drawn out. X-ray investigations showed that the chains are practically ideally oriented along the fibre axis. Periodic constancy of orientation along the fibre axis due to the relaxation of tensions is thought to be the reason for the formation of striped textures.

The characteristic peculiarities of thermotropic liquid-crystalline derivatives of cellulose when they are processed have been studied by Liang [115]. The degree of LC ordering in the molten state decreases in the order HOPC > EC > MC. Cellulose acetobutyrate and acetopropionate do not show LC properties when melted. Liang confirmed by polarization microscopy, scanning electron microscopy, wide-angle X-ray scattering, small-angle light scattering, double refraction and DSC that HOPC has a fibrillar structure during the process of shear flow. The orientation of rigid stick-like molecules in the amorphous regions has a very significant effect on the behaviour of the polymer during processing.

REFERENCES

1. Rogovin Z. A., Galbraikh L. S. *Chemical Transformations and Modifications of Cellulose*, Khimiya, Moscow **1979**.
2. Karlivan V. P. (ed.) *Methods of Investigation of Cellulose*, Zinatrie, Riga **1981**.
3. Rogovin Z. A. *Chemistry of Cellulose*, Khimiya, Moscow **1972**.
4. Bickles M., Segal L. (eds) *Cellulose and Its Derivatives*, Vol. I, Mir, Moscow **1974**.
5. Tarchevskii I. A., Marcheriko G. N. *Biosynthesis and Structure of Cellulose*, Nauka, Moscow **1985**.
6. Philipp B., Schleicher H., Wagenknecht W. *Chem. Technol.* **1977**, *7*, 702–709.
7. Philipp B., Schleicher H., Wagenknecht W. *Cellul. Chem. Technol.* **1975**, *9*, 265–82.
8. Philipp B., Schleicher H. in *III Int. Symp. on Chem. Fibres, Kalinin (USSR)*, **1981**, *2*, 260–78.
9. Mancier D., Vinendon M. *Bull. Soc. Chim. Fr.*, **1981**, *2*, 319–27.
10. Schweiger R. J. *Tappi.* **1974**, *57*, 86–90.

11. Philipp B., Schweiger R. *Plast. Kautsch.* **1983**, *30*, 65–69.
12. Myasoedova V. V. *Vysokomole. Soyedin. Krat. Soobshch.* **1988**, *30*, 666–68.
13. McCorsley C. C. *US Pat.* 4 246 221.
14. Litt V. Y., Kumar N. G. *US Pat.* 4 028 132.
15. Menault J., Rodier H. *Fr. Pat.* 2 358 435.
16. Adamova O. A., Myasoedova V. V., Krestov G. A. *A.c.SSSR* 1 164 237; publ. 30.06.85, B.I. N24.
17. Rodier H. *Fr. Pat.* 2 358 477.
18. Williams M. D. *US Pat.* 323 669.
19. Kulichikhin V. G., Belousov Yu. Ya., Platonov V. A. *All-Union Conference on Structure and Reactivity of Cellulose, Dissolution Mechanisms for Cellulose and Its Derivatives in Non-Aqueous Media*, BGU, Minsk **1982**, 27–31.
20. Yamaroto H. *Jpn. Pat.* 21 893.
21. Barton A. F. M. *Chem. Rev.* **1975**, *75*, 731–53.
22. Askadskii A. A., Matveyev Yu. I. *Dokl. Akad. Nauk SSSR* **1986**, *290*, 135–39.
23. Askadskii A. A., Matveyev Yu. I. *Chemical Structure and Physical Properties of Polymers*, Khimiya, Moscow **1983**.
24. Huyskens P. L., Haulait-Pirson M. C. *J. Coat. Technol.* **1985**, *55*, 57–67.
25. Hansen C. M. *J. Paint Technol.* **1967**, *104*, 39–45.
26. Myasoedova V. V. in *Abstracts of All-Union Conference on Chemistry and Physics of Cellulose*, NII KhTTs, Tashkent **1982**, 46.
27. Nikiforov M. Yu., Alper G. A., Durov V. A., Korolyov V. P., Vyugin A. I., Krestov G. A., Myasoedova V. V., Krestov A. G. *Solutions of Non-Electrolytes in Liquids*, Nauka, Moscow **1989**, 182–232.
28. Nesterov A. E., Lipatov Yu. S. *Phase State of Solutions and Mixtures of Polymers*, Naukova Dumka, Kiev **1987**.
29. Scott R. L. *J. Chem. Phys.* **1949**, *14*, 268–84.
30. Papkov S. P., Kulichikhin V. G. *Khim. Volok.* **1983**, 8–12.
31. Birshtein T. M., Merkuryeva A. A. *Vysokomol. Soyedin.* **1985**, *21*, 1208–16.
32. Valenti B., Sartirana M. L. *Nuovo Chim. D* **1984**, *3*, 73–89.
33. Shcheglov S. Yu., Klenin V. L., Frenkel S. Ya *Vysokomol. Soyedin.* **1986**, *27*, 1190–95.
34. Flory, P. J. *Proc. R. Soc. London, Ser. A* **1956**, *234*, 73–89.
35. Nies E., Konigsveld R., Kleintjens L. A. *Prog. Colloid. Polym. Sci.* **1985**, *71*, 2–14.
36. Kamide K., Saito M., Suzuki H. *Macromol. Chem. Rapid. Commun.* **1983**, *4*, 33–39.
37. Birshtein T. M., Merkuryeva A. A. *Vysomol. Soyedin.* **1985**, *27*, 1208–16.
38. Aharoni S. M. *J. Macromol. Sci.* **1982**, *21*, 287–98.
39. Gray D. G. in *Proceedings of the International Symposium on Fiber Science and Technology, Hakene*. **1985**, 17--18.
40. Pawlowski W. P., Gilbert R. D., Fornes R. E., Purrington S. T. *J. Polym. Sci.* **1988**, *26*, 1101–10.
41. Flory P. J., Ronca C. *Mol. Cryst. Liq. Cryst.* **1979**, *54*, 289–310.
42. Matheson R. R., Flory P. J. *J. Macromol.* **1981**, *14*, 954–60.
43. Bosch A. T., Maissa J., Sixou P. *Phys. Lett. A* **1983**, *94*, 113–23.
44. Bosch A. T., Maissa J., Sixou P. *Nuovo Chim. D* **1984**, *3*, 95–103.
45. Saito M. *Polym. J.* **1983**, *15*, 213–23.
46. Cotter M., Wacker *D. Phys. Rev. A.* **1978**, *18*, 2669–73.
47. Gilbert R. D. *Compos. Syst. Nat. Synth. Polym. Amsterdam* **1986**, 97–103.
48. Usoltseva V. A. *Zh. Vses. Khim. Ova.* **1983**, *2*, 122–31.
49. Usoltseva N. V. *Zh. Vses. Khim. Ova.* **1983**, *2*, 156–65.
50. Plate N. A. (ed.) *Liquid-Crystalline Polymers* Khimiya, Moscow **1988**.
51. Metzner A. B., Prilutski G. M. *Polym. Liq. Cryst.* **1987**, *10*, 661–91.

52. Morachevskii A. G., Lilich L. S. (eds) *Chemistry and Thermodynamics of Solutions* LGU, Leningrad **1986**.
53. Usoltseva N. V. *Izv. Akad. Nauk SSSR. Ser. Fiz.* **1989**, *53*, 1992–2003.
54. Patel D. L., Gilbert R. D. *J. Polym. Sci., Polym. Phys. Ed.* **1981**, *19*, 1231–36.
55. Conio G., Bianchi B., Tealdi A. in *Abstracts of 5th Conference of Socialist Countries on Liquid Crystals, Odessa (USSR)* **1983**, *2(1)*, 147–48.
56. Gilbert R. D., Patton P. A. *Prog. Polym. Sci.* **1983**, *9*, 115–31.
57. Takahashi J. *J. Text. Mach. Soc. Jpn.* **1985**, *38*, 217–25.
58. Shoul M., Aharoni S.M. *Mol. Cryst. Liq. Cryst.* **1980**, *56*, 237–41.
59. Kulichikhin V. G., Golova L. K. *Khim. Dreves.* **1985**, *3*, 9–27.
60. Nishio Y., Roy S. K., St John Manley R. *Polymer* **1987**, *28*, 1385–90.
61. Weigel P., Hirte R., Zenke D. *Acta Polym.* **1984**, *35*, 83–88.
62. Krestov G. A., Myasoedova V. V., Belov S. Yu., Alexeyeva O. V. *Dokl. Akad. Nauk SSSR* **1987**, *293*, 174–76.
63. Suto N. *Kaigai Kobunsi Kenku* **1987**, *33(6)*, 115–16.
64. Sato T., Teramoto A. *High Polym. Jpn.* **1988**, *37*, 278–81.
65. Navard P., Haudin J. M. in *Abstracts of IUPAC MACRO'83, Budapest* **1983**, 409–12.
66. Chanzy H., Dube M. *J. Polym. Sci., Polym. Lett. Ed.* **1979**, *17*, 219–26.
67. Pozhkova O. V., Myasoedova V. V., Krestov G. A. *Khim. Dreves.* **1985**, *2*, 26–29.
68. Petropavlovskii G. A., Pogodina T. E., Shek V. N. *Cellul. Chem. Technol.* **1986**, 20, 3–10.
69. Chanzy H., Peguy A. *J. Polym. Sci., Polym. Phys. Ed.* **1980**, *19*, 1137–44.
70. Alexeyeva O. V., Myasoedova V. V., Krestov G. A. *Russ. J. Appl. Chem.* **1987**, *60*, 2523–26.
71. Hawkinson D. E., Gilbert R. D., Fornes R. E., Kohout E. *Polym. Mater. Sci. Eng.* **1987**, *57*, 947.
72. Iovleva M. M. in *Investigation Methods for Cellulose* (Karlivan V. P. ed.) Zinatne, Riga **1981**, 138–47.
73. Malafeyeva I. M., Averyanova V. M. *Processes of Gel Formation in Polymer Systems* Izdvo Srat. un-ta, Saratov **1977**, 43–48.
74. Derzhavina E. D., Zakurdayeva N. P. *Problems of Technology of Photo Materials and Chemicals for Their Production*, Moscow **1987**, 114–22.
75. Andreyeva V. M., Tsilipotkina M. V., Tager A. A., Safronova V. A., Shilnikova N. I. *Vysokomol. Soyedin.* **1986**, *28*, 2147–51.
76. Abaturova A. A., Vlodavets I. A., Rebinder P. A. *Kolloid. Zh.* **1972**, *24*, 315–19.
77. Werbowij R. S., Gray D. G. *Mol. Cryst. Liq. Cryst.* **1976**, *34*, 97–103.
78. Luise R. R., Morgan R. D., Panar M. in *Colloid and Surface Science Symposium, University of Missouri, Rolla (MO)*, **1979**, Abstr. Par. 53.
79. Tsvetkov V. N. *Rigid-Chain Polymer Molecules*, Nauka, Leningrad **1986**.
80. Meeten G. H., Navard P. *Polymer* **1983**, *24*, 815–19.
81. Alexeyeva O. V., Myasoedova V. V., Krestov G. A. *Russ. J. Phys. Chem.* **1986**, *60*, 2441–45.
82. Navard P., Haudin J. M. *Am. Chem. Soc. Polym. Prepr.* **1983**, *24*, 267–68.
83. Bheda J., Fellers J. F., White J. L. *Colloid Polym. Sci.* **1980**, *258*, 1335–42.
84. Bianchi E., Ciferri A., Conio G. *Macromolecules* **1986**, *19*, 630–36.
85. Patel D. L., Gilbert R. D. *J. Polym. Sci., Polym. Phys. Ed.* **1982**, *20*, 1019–28.
86. Yunusov B. Yu., Khanchich O. A., Dibrova A. K. *Vysokomol. Soyedin. Krat. Soobshch.* **1982**, *24*, 414–18.
87. Alexeyeva O. V. *PhD Thesis* University of Ivanovo **1987**.
88. Brestkin Yu. V., Volkova L. A., Kutsenko L. I. *Vysokomol. Soyedin. Krat. Soobshch.* **1986**, *28*, 32–37.

89. Volkova L. A., Kutsenko L. I., Kulakova O. M., Meltser Yu. A. *Vysokomol. Soyedin. Krat. Soobshch.* **1986**, *28*, 27–31.
90. Dayan S., Maissa P., Vellutini M. J. *J. Polym. Sci., Polym. Lett. Ed.* **1982**, *20*, 33–43.
91. Khokhlov A. R., Semenov A. M. *Physica A* **1982**, *112*, 605–14.
92. Viney C., Windle A. H. *Mol. Cryst. Liq. Cryst.* **1987**, *148*, 145–61.
93. Gilbert R. D. *Compos. Syst. Natur. Polym., Amsterdam* **1986**, 97–106.
94. Werbowij R. S., Gray D. G. *Macromolecules* **1980**, *13*, 69–73.
95. Donald A. M., Viney C., Ritter A. P. *Liq. Cryst.* **1986**, *19*, 287–300.
96. Werbowij R. S., Gray D. G. *Liq. Cryst.* **1984**, *17*, 1512–20.
97. Conio G., Bianchi E., Ciferri A. *Liq. Cryst.* **1983**, *16*, 1264–71.
98. Bianchi E., Ciferri A., Conio G. *Liq. Cryst.* **1986**, *19*, 630–36.
99. Friend F., Sixou P. *J. Polym. Sci., Polym. Chem. Ed.* **1984**, *22*, 239–47.
100. Nishio J., Susuki S., Takanashi T. *Polym. J.* **1985**, *17*, 753–60.
101. Zugenmaier P. in *Proceedings of the International Symposium on Fiber Science and Technology, Hakene* **1985**, 279.
102. Huang Y., Chen M.-C., Li L.-S. *Acta Chim. Sin.* **1988**, *46*, 367–71.
103. Kamide K., Obajama K., Matsui T. *Carbohydr. Res* **1986**, *16*, 273–76.
104. Ambrosino S., Khallala T., Seurin M. J., Ten Bosch A., Friend F., Maissa P., Sixou P. *J. Polym. Sci.* **1987**, *25*, 351–57.
105. Laivins G. V., Sixou P. *J. Polym. Sci. B* **1988**, *26*, 113–25.
106. Gray D. G. in *Proceedings of the International Symposium on Fiber Science and Technology, Hakene* **1985**, 17–18.
107. Shimamura K., White J., Feller J. *J. Appl. Polym. Sci.* **1981**, *26*, 2165–87.
108. Jiang B. in *3rd Annu. Meet. Polym. Process Soc., Stuttgart, 1987, Program and Abstr.* **1987**, 152.
109. Nishio Y., Takanashi T. in *Proceedings of the International Symposium on Fiber Science and Technology, Hakene* **1985**, 140.
110. Shibayev V. P., Yekayeva I. V. *Vysokomol. Soyedin.* **1987**, *29*, 2647–53.
111. Pawlowski W. P., Gilbert R. D., Fornes R. E., Parrington S. T. *J. Polym. Sci. B* **1987**, *25*, 2293–2301.
112. Lawins G. V., Gray D. G. *Macromolecules* **1985**, *18*, 1746–52.
113. Navard P., Zacharades A. E. *J. Polym. Sci. B* **1987**, *25*, 1089–98.
114. Pawlowski W. P., Gilbert R. D., Fornes R. E., Parrington S. T. *J. Polym. Sci. B* **1988**, *26*, 1101–10.
115. Liang B. in *3rd Annu. Meet. Polym. Process Soc., Stuttgart, 1987, Program and Abstr.* **1987**, 52.
116. Nishio Y., St John Manley R. *Polym. Eng. Sci* **1990**, *30*, 71–82.
117. Nishio Y., Haratani T., Takahashi T. *Macromolecules* **1989**, *22*, 2547–49.
118. Masson J.-F., St John Manley R. *Macromolecules* **1991**, *24*, 6670–79.
119. Funaki Y., Ueda K., Saka S., Soejima S. *J. Appl. Polym. Sci.* **1993**, *48*, 419–24.
120. Tang L.-G., Hon, D. N.-S. Zhu Y.-Q. *J. Appl. Polym. Sci.* **1997**, *64*, 1953–60.
121. Nishio Y., St John Manley R. *Macromolecules* **1988**, *21*, 1270–77.
122. Nishio Y., Roy S. K., St John Manley R. *Polymer* **1987**, *28*, 1385–90.
123. Zhang, Z. B., McCormick J. *Appl. Polym. Sci.* **1997**, *66*, 293–305.
124. Davè V., Tamagno M., Focher B., Marsano E. *Macromolecules* **1995**, *28*, 3531–39.

CHAPTER 2

Influence of the Solvent on the Equilibrium and Kinetic Rigidity of the Molecular Chain of Cellulose and Its Derivatives in Solution

Thermodynamic flexibility is a fundamental equilibrium property of macromolecules of cellulose and its derivatives. It is the ability of the polymer chains to assume numerous spatial forms or conformations. This ability is due to the existence of internal degrees of freedom related to internal rotation about bonds in the main polymer chain.

The theory of flexible chains of synthetic polymer molecules in dilute solution is based on extensive experimental material and is very well developed. In contrast, the data for the estimation of conformational and hydrodynamic parameters in solutions of cellulose are often ambiguous. Interest in rigid-chain derivatives of cellulose has developed recently. This has been discussed by Krestov *et al.* [1] and by Gotlib *et al.* [2].

The rigidity of the cellulose chain can be changed by solvation with non-aqueous solvents and by the formation of polymer–solvent complexes [3]. Depending on the specificity of the complex formed, either an increase or decrease in chain rigidity can occur.

The estimation of polymer chain rigidity in solution is important for both the theory and the practice of polymer science. In particular, the phase state and rheological characteristics of a system play a considerable role in the production of fibres from spinning solutions, films and thickening agents for printing dyes.

If a system is in the liquid-crystalline state it is possible to produce items with improved physico-mechanical and unique optical properties. However, in the case of many cellulose derivatives, owing to insufficient rigidity of the polymer macromolecules, it is either impossible to obtain the LC state or it can only be achieved at very high polymer concentrations.

Highly concentrated solutions are, however, unsuitable for technological processing because of extremely high viscosity. In some cases, the selection of an appropriate solvent can help to achieve an increase in macromolecular rigidity and a simultaneous decrease in viscosity. This greatly increases the possibilities for improving and producing products made from cellulose derivatives.

1 MODELLING OF THE CONFORMATIONAL CHARACTERISTICS OF MACROMOLECULES OF CELLULOSE AND ITS DERIVATIVES

The tendency towards thermotropic and lyotropic mesomorphism by polymers and the nature of their mesogeneity is determined to a considerable extent by the degree of anisodiametry. In this connection, the analysis of the forms which macromolecules can take up and of the influence of different physico-chemical parameters on these forms gives an insight into ways of attaining the lyotropic LC state.

This analysis involves finding the interrelation between the parameters characterizing the size of a molecule (distance between the ends of polymer chain r, radius of gyration s) and molecular characteristics (molecular mass M and contour length L). This problem can be tackled by computer modelling of the mechanism of chain molecule flexibility. These models can be based on the free-conjugated, the rotary-isomeric and the persistent mechanisms of flexibility [4].

In the free-conjugated chain model that was developed by Kuhn [5] and Guth and Mark [6], the real chain is substituted by an 'equivalent' one, consisting of N linear segments of length A, the spatial orientations of which are mutually independent. The condition of 'equivalence' assumes the selection of N and A values ($L=NA$) at which the linear dimensions of a model free-conjugated chain correspond to the dimensions of a real chain molecule. The length of the segment into which the chain is divided serves here as a measure of the equilibrium rigidity of the chain and is called the Kuhn segment.

Owing to internal thermal motion, macromolecules in solution can assume an infinite number of 'momentary' conformations [7]. The average square distance between the chain ends $\langle r^2 \rangle$ and average square radius of gyration $\langle s^2 \rangle$ are the characteristics of the statistical dimensions of a molecule. Linear statistical dimensions of a free-conjugated chain are related to the length of a Kuhn segment as follows:

$$\langle r^2 \rangle = LA; \ \langle s^2 \rangle = (LA)/6 \tag{2.1}$$

Further development of models of the conformational state of polymer chains was achieved by attempts at improving the agreement between the flexibility mechanism and the structural–geometric parameters of macromolecules, i.e. virtual angle values and valent bond lengths. The simplest variant of such a model was put forward by Eyring [8]. This was a chain consisting of an arbitrary number, n, of links with length l and rotating with respect to each other through bond angle

$\pi - \theta$. For a long chain ($n \to \infty$):

$$\langle r^2 \rangle = nl^2(1 + \cos\theta)/(1 - \cos\theta) \tag{2.2}$$

$$\langle s^2 \rangle = [(nl^2)/6](1 + \cos\theta)/(1 - \cos\theta) \tag{2.3}$$

It follows from Eqs (2.2) and (2.3) that if the chain is long enough and has unhindered rotation at fixed bond angle, then it becomes Gaussian (i.e. equivalent to freely jointed), since its linear dimensions grow in proportion to the contour length of the chain (nl). It follows that one can apply the statistics of freely jointed chains to real polymer molecules of considerable length and use the Kuhn segment length as a measure of their rigidity in solution.

The approach postulating deformability of bond angles and bonds, while being true for flexible chain polymers, is inapplicable to semi-rigid and rigid chain polymers which include cellulose and its derivatives. It is well established that molecules of polymers possessing high equilibrium rigidity have the form of a Gaussian tangle in solution if they are sufficiently long [9]. It is not possible to visualize in such a form the macromolecules of polypeptides [10], cellulose derivatives [11] and desoxyribonucleic acid [12] which contain only a small number of Kuhn segments in the chain. The assumption that bond angles (and sometimes bonds, especially hydrogen bonds) are deformed is not applicable for polymer molecules with increased equilibrium rigidity [9]. Simulation of the flexibility mechanism, taking deformability into account, can be achieved in the persistent or worm-like chain model [13]. This model takes into account the orientational short-range action of elements which constitute the chain. In this case, the spatial orientations of the neighbouring elements of the chain are not mutually independent. The direction of the first element is, to some extent, communicated along the chain. This is in contrast to the freely jointed chain in which the orientation of the segments is limited by bond angles. In the case of a continuous worm-like chain ($l \to \infty$), the curvature is constant along its entire length and determined by the length of persistence q. The size of a worm-like chain is given by Porod's equation [14]:

$$\langle r^2 \rangle = 2q^2(L/q - 1 + \mathrm{e}^{-L/q}) \tag{2.4}$$

If a persistent chain is long enough, its dimensions correspond to a Gaussian distribution. The persistent length, q, is equal to half the length of the Kuhn segment, A, of the equivalent Gaussian chain. In the case, of a short worm-like chain its form approximates to stick-like. In this case, $\langle r^2 \rangle = L^2$.

There is an advantage in applying the model of a worm-like chain to macromolecules of cellulose and its derivatives. It does not seem to be necessary to make *a priori* assumptions about the flexibility mechanism because of the constancy of bonds and bond angles. It is possible to accommodate within the framework of the model the entire range of forms which macromolecules assume is solution, from rod-like to statistical tangle conformation.

Novahovskaya and Strelena [15] reported Bogdanetskii's analysis of the values of the temperature coefficient of the intrinsic viscosity, $\mathrm{dln}[\eta]/dT$ for rigid chain polymers, using the model of a worm-like cylinder described by Yamakawa and Fujii [16]. This is reflected in the value of $\mathrm{dln}(\langle s^2 \rangle \mathrm{d}M)_\infty/\mathrm{d}T$. The data for this parameter and for the chain rigidity of ethylcellulose, cellulose nitrate, cellulose tributryrate and cellulose tricarbanilate have been tabulated [16].

2 HYDRODYNAMIC AND CONFORMATIONAL PARAMETERS OF CELLULOSE AND ITS DERIVATIVES IN NON-AQUEOUS SOLUTIONS

There are several experimental methods for obtaining information on the conformational structure and rigidity of a macromolecular chain. The translatory and rotary friction are the most important phenomena which reflect the peculiarities of the structure of polymer molecules. These two phenomena determine the rate of diffusion, rate of sedimentation and the viscosity of polymer solutions. Quantitative interpretation on the molecular level of experimental data is possible if the relationship between conformational characteristics of molecules and their hydrodynamic properties in solution can be assumed. The polymer molecule under investigation can be simulated as a geometric body having a certain configuration. The translatory and rotary motion is described by applying the laws of the hydromechanics of macroscopic bodies in viscous medium. The worm-like cylinder, that is, the cylinder curved so that the form of its axis line can be described by the equation of the worm-like chain, is the model which reflects the most adequately the hydrodynamic behaviour of a molecule of a rigid chain polymer in solution [17–20].

The prediction of the intrinsic viscosity, $[\eta]$, of a solution of rigid-chain molecules on the basis of the model of the worm-like cylinder has been developed by Yamakawa and co-workers. The first calculations [16] of $[\eta]$ were carried out without taking edge-effects into account. In a later paper, Yamakawa and Yoshizaki [21] made allowance for the semi-spherical form of the ends of the worm-like cylinder and obtained the following expression for the region $L/A > 2.278$:

$$M/[\eta] = \Phi_\infty^{-1}(M_L/A)^{3/2}M^{1/2}\Psi(L, d, A) \qquad (2.5)$$

where $M_L = M/L$, d is the diameter of the polymer, Φ_∞ is the viscometric constant at infinite dilution with a value of 2.87×10^{23} mol^{-1} and

$$\Psi = 1 - [C_1(A/L)^{1/2} + C_2(A/L) + C_3(A/L)^{3/2} + C_4(A/L)^2] \qquad (2.6)$$

Coefficients C_1, C_2, C_3 and C_4 depend on d/A and are represented by a polynomial series published elsewhere [20]. The introduction of a complex functional dependence of Ψ on L, d and A is due to the necessity for taking into account the influence of the effects of excluded volume and 'leakage' of macromolecules on their hydrodynamic properties.

'Leakage' of the chain is the resistence experienced by the chain during its motion in a solvent. This is equal to the sum of the resistances which would be experienced by all its elements if each of them moved in the absence of the rest. In this sense, the completely extended chain possesses the greatest 'leakage'. When the chain molecule 'rolls up' into a tangle, there is screening of the segments that are situated far from its surface, which is equivalent to the decrease in the hydrodynamic interaction of the molecule with solvent. The degree of 'rolling up' of the chain depends on its equilibrium rigidity, length and thermodynamic quality of solvent. For macromolecules of cellulose and its derivatives, the conformation in the region of molecular masses $\sim 10^5$–10^6 g mol^{-1} is different from a tangle-like conformation. As a result, 'leakage' exerts a considerable influence on their viscometric behaviour.

Excluded volume effects arise from a limitation of the number of mutual orientations of chain elements due to their mutual non-overlapping. This causes an increase of the linear dimensions of the macromolecular tangle:

$$\langle r^2 \rangle^{1/2} = \alpha \langle r^2 \rangle_\theta^{1/2} \tag{2.7}$$

where α is the degree of swelling of the tangle and $\langle h^2 \rangle_\theta^{1/2}$ is its linear dimension in the absence of excluded volume effects. The influence of the effect of excluded volume on the hydrodynamics of molecules becomes stronger with increase in the length and decrease in the equilibrium rigidity of macromolecules. When the macrochain rolls up into a tangle, the number of intramolecular contacts increases and volume effects become stronger.

The selection of θ conditions (conditions under which the volume effects are compensated by the mutual attraction between the chain elements) for many polymers is difficult [22]. It is simpler to estimate the equilibrium rigidity of a chain, which is not perturbed by the volume effects, by investigating the hydrodynamic properties of low-molecular-mass fractions of polymers. The Mark–Kuhn equation for a non-perturbed chain under θ conditions:

$$[\eta]_\theta = K_\theta M^{1/2} \tag{2.8}$$

predicts the linear increase in the intrinsic viscosity with increase in $M^{1/2}$, where M is the relative molecular mass. This may be written as

$$[\eta]_\theta / M^{1/2} = K_\theta = \Phi_\infty (LA/M)^{3/2} \tag{2.9}$$

Non-ideality of solvent results in a greater increase in the dimensions of a macromolecule with increase in its molecular mass. Stockmayer and Fixman [23] published an equation which takes into account the dimensions of the perturbed chain:

$$[\eta]_\theta/M^{1/2} = \Phi_\infty(LA/M)^{3/2} + 0.51\Phi_\infty BM^{1/2} \quad (2.10)$$

Extrapolation of a plot of experimental values of $[\eta]/M^{1/2}$ against $M^{1/2}$ to $M^{1/2} = 0$ enables the value of $A/M_L^{3/2}$ to be obtained. The value of the parameter A of a chain, which is not perturbed by volume effects, can then be calculated.

The applicability of the Stockmayer–Fixman method of estimation of the non-perturbed length of a Kuhn segment has been confirmed by Tsvetkov [9]. The limiting value of $[\theta]/M^{1/2}$ at zero value of $M^{1/2}$ for poly(vinyl cinnamate) in a good solvent is equal to the value of $[\theta]/M^{1/2}$ obtained by direct measurement in a θ solvent.

With increasing equilibrium rigidity of a chain the influence of volume effects on the macromolecular sizes becomes weaker: a smaller number of intramolecular contacts corresponds to a smaller degree of 'rolling up'. This is confirmed by experimental measurements. Tsvetkov put forward the hypothesis that the role of the equilibrium rigidity of a chain was more important than volume effects in determining conformational properties [10]. As evidence he cites data for negative temperature coefficients of the intrinsic viscosity of solutions of cellulose derivatives.

Application of the theoretical Eq. (2.5) to estimate the non-perturbed dimensions of molecules of rigid chain polymers is extremely difficult because the dependence of $M/[\eta]$ on $(L/A)^{1/2}$ is non-monotonic. It passes through a minimum at small values of L. This is associated with the deviation of the macromolecular conformation from the tangle-like conformation. With increasing L/A, the dependence represented by Eq. (2.5) degenerates into an asymptotic function which, with good approximation, can be described by the following equation [24]:

$$M/[\eta] = \Phi_\infty(M_L/A)^{3/2}M^{1/2} + 2.2\Phi_\infty^{-1}(M_L^2/A)[l(A/d) - 0.755] \quad (2.11)$$

which is true for long worm-like spherocylinders ($L/A > 20$).

If the value of M_L is known then the size of a Kuhn segment may be found by plotting $M/[\eta]$ against $M^{1/2}$ and finding the slope of the curve at a particular value of $M^{1/2}$. This follows from the relation

$$\partial(M/[\eta])/\partial M^{1/2} = (M_L/A)^{3/2}\Phi_\infty^{-1} \quad (2.12)$$

The hydrodynamic diameter, d, of the macromolecule can be found from the value of A and the intercept on the ordinate.

High equilibrium rigidity of macromolecules limits the applicability of the extrapolation Eq. (2.11) to the region of high-molecular-mass fractions of polymers.

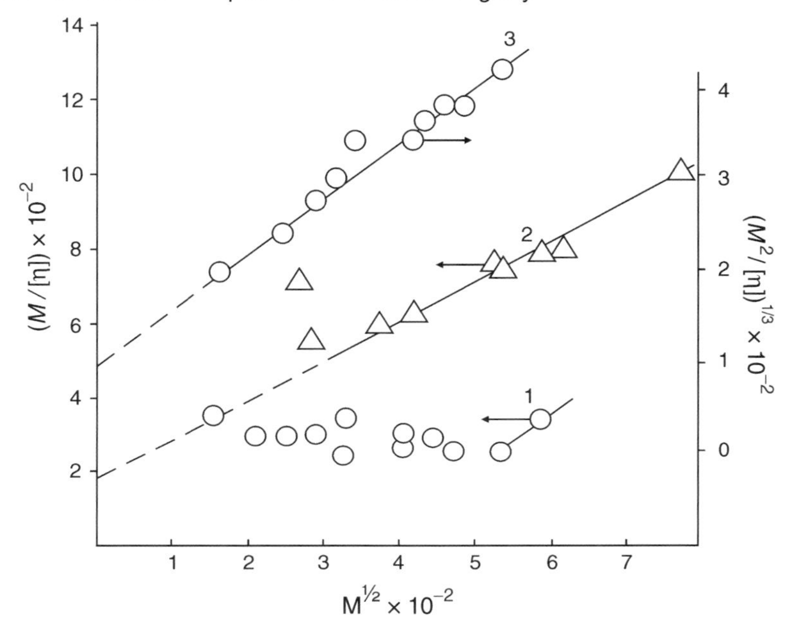

Figure 2.1 Dependence of $M/[\eta]$ (1, 2) and of $M^2/[\eta]^{1/3}$ (3) on $M^{1/2}$ for solutions of cellulose nitrate fractions [9]. 1, 3, A sample with a degree of substitution of 2.4 in ethyl acetate (No. 15 in Table 2.2); 2, a sample with a degree of substitution of 1.14 in DMAA containing 6% LiCl (No. 22 in Table 2.2)

With increasing skeletal rigidity of the chain, the deviation from the properties of Gaussian tangles becomes pronounced at high molecular masses.

It follows that there is a greater deviation from a linear dependence of $M/[\eta]$ upon $M^{1/2}$ at higher values of M. Figure 2.1 shows plots of the relationship $M/[\eta] = f(M^{1/2})$ for fractions of cellulose nitrate with degrees of substitution (DS) of 2.4 and 1.14. The nitrate with the lower value of DS possesses lower equilibrium rigidity (see Table 2.2, later). In the case of this material noticeable deviations from linearity begin at $M^{1/2} > 400$.

The sample with the higher equilibrium rigidity did not follow a linear relationship corresponding to Eq. (2.11) over any part of range of values of M which were investigated. However, there is an analytical expression in the form of a simple linear dependence between the viscometric data and molecular mass which is applicable over a wider region of L/A values [25], i.e.

$$(M^2/[\eta])^{1/3} = \Phi_\infty^{-1/3}(M_L/A)^{1/2}M^{1/2} + (k/4.65\,A_0) \times (M_L/3\pi)(l_n[A/d] - 1.065) \qquad (2.13)$$

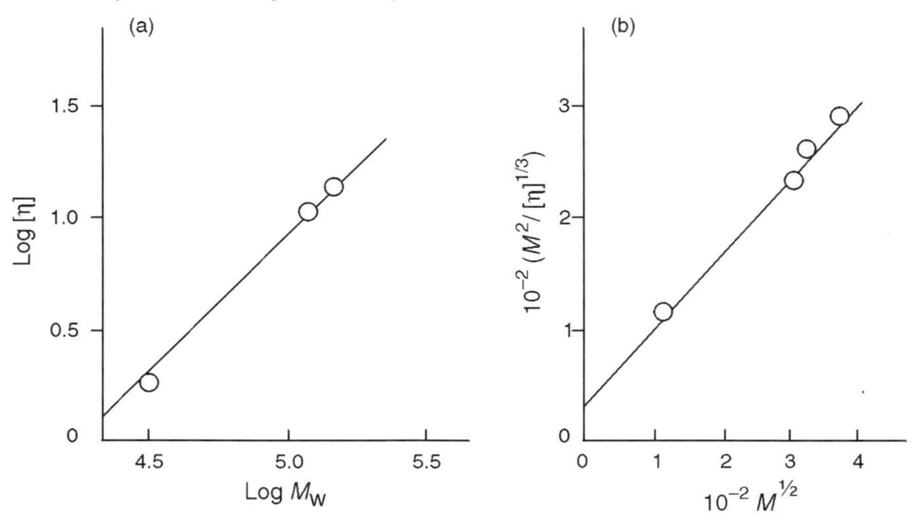

Figure 2.2 Solutions of cellulose in a mixture of CF_3COOH and $C_2H_4Cl_2$ ($x_{C_2H_4Cl_2} = 0.3$). (a) Dependence of $\log[\eta]$ on $\log M_W$; (b) Dependence of $M^2[\eta]^{1/3}$ on $M^{1/2}$

where A_0 is the hydrodynamic invariant for a given polymer–solvent system and k is Boltzmann's constant.

Equation (2.13) indicates that there is a linear dependence of the form $M^2/[\eta] = f(M^{1/2})$ over a wide range of molecular masses. This is illustrated in Figure 2.2. The equilibrium rigidity of the macrochain parameter, A, can be determined from the slope of the straight line if M_L is known. This follows because the slope of the line is given by

$$\partial(M^2/[\eta])^{1/3}/\partial M^{1/2} = (ML/A)^{1/2}\Phi_{\infty}^{-1/3} \tag{2.14}$$

where $\Phi_{\infty} = 2.87 \times 10^{23}$ mol^{-1}. The diameter of the chain, d, can be calculated from the intercept of the line on the ordinate axis.

These methods are only applicable to polymers with very high chain rigidity. Cellulose nitrates, for which $\lambda = 5.1 \pm 0.1$ Å [10, 21], meet this criterion better than other cellulose derivatives. In practice this value of λ has been used in calculations involving other cellulose derivatives [26, 27].

Determination of λ or M_L from Eqs (2.11) and (2.13) is acceptable if there is an independent experimental determination of A or α. One method of doing this has been described by Bohdanecky and Kovar [28]. This is based on the assumption of the equality between the hydrodynamic volume occupied by 1 g of the worm-like chain and the partial specific volume of polymer, $\bar{V}$. A series of empirical correlations were used to calculate d [Eqs (11)–(13) in Ref. 27]. If the value of d is known then the slope and intercept of a plot of $(M^2/[\eta])^{1/3}$ against $M^{1/2}$ enable A and M to be calculated.

This procedure has been used to determine the hydrodynamic and conformational parameters in a non-aqueous solution of cellulose in mixtures of trifluoroacetic acid with chlorinated hydrocarbons [1]. The relation between molecular mass and intrinsic viscosity was determined experimentally [29] using cellulose samples with degrees of polymerization from 150 to 910 as measured by sedimentation methods using an analytical ultracentrifuge. Solvents used were mixtures of trifluoroacetic acid with chloroform, methylene chloride or 1,2-dichloroethane. Viscosities at concentrations from 0.8 to 2.5 g dm^{-3} were measured with a capillary viscometer and values of the intrinsic viscosity were determined graphically.

Non-destructuring solvents were selected in order to determine reliable molecular mass characteristics of cotton and wood celluloses. The optimum mole fraction of chlorinated hydrocarbon was 0.3 in mixtures of CF_3COOH and $CHCl_3$ or CH_2Cl_2 or $C_2H_4Cl_2$. This was found by analysis of data on cellulose solubility and acid self-association in non-aqueous solvents [30].

Sedimentation analysis of cellulose solutions in CF_3COOH and its mixtures with chlorinated hydrocarbons indicated a considerable influence of the nature and composition of solvents on the form of sedimentograms.

In the case of cellulose solutions in 70 mol% CF_3COOH + 30 mol% $C_2H_4Cl_2$ or 30 mol% $CHCl_3$, the form of the sedimentograms is traditional for dilute solutions of polymers. The forms of sedimentograms of the solutions in CF_3COOH and CF_3COOH–CH_2Cl_2 are of multimodal character and their reproduction is difficult. The average molecular mass was calculated by extrapolation to infinite dilution of the effective molecular mass using the following expression [30]:

$$(1/\bar{M})_{corr} = 1/\bar{M}_W + A_2C + \cdots \tag{2.15}$$

Values of $\bar{M}_W$ were corrected for polydispersity, taking into account the polydispersity coefficient according to the equation

$$\bar{M}_{W\,corr} = \bar{M}_W(a+1)\Gamma(a+1)/K \tag{2.16}$$

where a is the coefficient of the Mark–Kuhn–Houwink equation, Γ is the gamma function and $K=2/7$.

The specific partial volume $\bar{V}$, which is necessary for the calculation of the average molecular mass from the sedimentation data, was determined picnometrically. The values of $\bar{V}$ and ρ at 298 K were 0.586 cm^3 g^{-1} and 1.393 g cm^{-3}, respectively.

Values of molecular mass and intrinsic viscosities are given in Table 2.1.

The viscosity of cellulose solutions in CF_3COOH decreases owing to break-up of the cellulose [31]. The monitoring of cellulose (DP 600 and 900) solutions in the

Table 2.1 Molecular masses and intrinsic viscosities of cellulose samples having various degrees of polymerization (DP)

DP of cellulose	$M \times 10^{-3}$	$[\eta]$/dl g^{-1}	$A \times 10^4$/cm^3 g^{-1} mol^{-1}
150	24.3 [a]		
150	21.0	1.86	36
600	97.2 [a]		
600	91.0	7.84	31
755	122.0 [a]		
750	140.0	8.81	27
810	147.0 [a]		
810	165.0	9.81	24

[a] Values of masses of cellulose samples as shown on labels

$CF_3COOH{-}C_2H_4Cl_2$ mixture showed that the viscosity did not change over a period of 2 months, within the limits of experimental error. This indicates that no destruction of cellulose occurred.

The measurements and calculations described above provided data for the equations $[\eta] = f(\bar{M}_W)$ and $(M^2/[\eta])^{1/3} = f(M^{1/2})$ and allowed the determination of the constants of the Mark–Kuhn–Houwink equation by a least-squares method.

The model of a persistent chain was used to estimate non-perturbed dimensions of cellulose macromolecules in a non-aqueous solvent. The dependence in Figure 2.2b is well described by the equation $(M^2/[\eta])^{1/3} = f(M^{1/2})$.

The intercept on the ordinate $A_\eta = 34$, and the slope of straight line, $B_\eta = 0.65$, were used to calculate the values of Kuhn segment (A), diameter of macromolecule (d) and other conformational parameters in accord with the following equations [27]:

$$d/A_0 = 0.5B_\eta^4\bar{V}/A \tag{2.17}$$

$$\log d = -0.173 + \log(d^2/A_0) \tag{2.18}$$

$$A_0 = 0.46 - 0.53\log d \tag{2.19}$$

$$M_L = A_\eta\Phi_\infty^{1/3} \tag{2.20}$$

$$h^2/M = (N_\eta\Phi_\infty^{1/3})^{-2};\ A = r^2M_L/M \tag{2.21}$$

The value of the conformational parameters are $A = 170 \times 10^{-8}$ cm, and $d = 5.5 \times 10^{-8}$ cm.

The closeness of the parameter η to unity, the size of the Kuhn segment and the absolute values of the second virial coefficients given in Table 2.1 indicate that cellulose in this solution is a semi-rigid-chain polymer. The repeated structural unit of cellulose is shown elsewhere [29]. The contour length λ of one link of cellulose, i.e. the projection of the lengths of bonds on to the chain axis, equals 5.33 Å; the molecular mass of the link is $M = 162$.

The method of calculating A was proposed by Papkov [32]. He postulated that the concentration C^*, at which a sharp increase in the slope of a plot of η_{spec}/C against C occurs, is the concentration at which the distances between centres of inertia of macromolecules become comparable to the sizes of the macromolecules themselves.

This hypothesis allows the estimation of the diameter of spheres of rotation of macromolecules, d, from the value of C^*:

$$d \approx (2.5M/C^*)^{1/3} \times 10 \tag{2.22}$$

where d is the diameter of rotation sphere (Å) and C^* is the upper concentration boundary of diluted solution (g per 100 cm^3).

For rigid-chain polymers, Papkov made the assumption $d \approx \langle r^2 \rangle$. The equilibrium rigidity of macromolecules, $A = 2q$, can then be calculated using Porod's Eq. (2.4). Further relationships involving experimental values of $[\eta]$ [Eqs (2.11) and (2.13)] allow the calculation of d and M.

It is now possible to calculate molecular parameters by a technique which involves comparison of experimental and theoretically calculated values of intrinsic viscosities. The equation published by Yamakawa [Eq. (2.5)] relates $[\eta]$ with the molecular characteristics of a polymer (M_L, A, d and L). Values of these parameters can be continuously varied in an appropriate computer program until the corresponding value of the intrinsic viscosity calculated from the equation agrees with the experimental value. It can then be assumed that the values of M_L, A, d and L obtained by the computer correspond to the actual conformation of the macromolecule. This method of calculating these parameters eliminates the assumptions that were made when Eq. (2.5), which has a theoretical basis, was substituted by the approximate expressions (2.11) and (2.13).

Jerbojevich *et al.* [26] studied the equilibrium rigidity of cellulose in *N,N*-dimethylacetamide–LiCl mixture. They assumed the length of the projection of a monomeric link of glucopyranose chain to be 5.14 Å. Variation of the value of the hydrodynamic diameter from 4.5 to 6 Å and of the persistent length, q, from 70 to 140 Å showed that the worm-like cylinder with $q = 110 \pm 10$ Å was the best model to fit the behaviour of real cellulose macromolecules in solution (Figure 2.3).

Similar procedures for calculating the conformational parameters are described in other papers [17, 33]. Tsvetkov *et al.* [34] extended the number of parameters during the calculations and included the contour lengths of chains, A. The values obtained agree satisfactorily with values obtained by extrapolation of a plot of $(M^2/[\eta])^{1/3}$ against $M^{1/2}$ to a value of $M = 0$.

There are limitations to the use of the variation method of calculating A and the conformational parameters. This is due to the impossibility of determining their independent values. Numerous pairs of d and A values can correspond to the

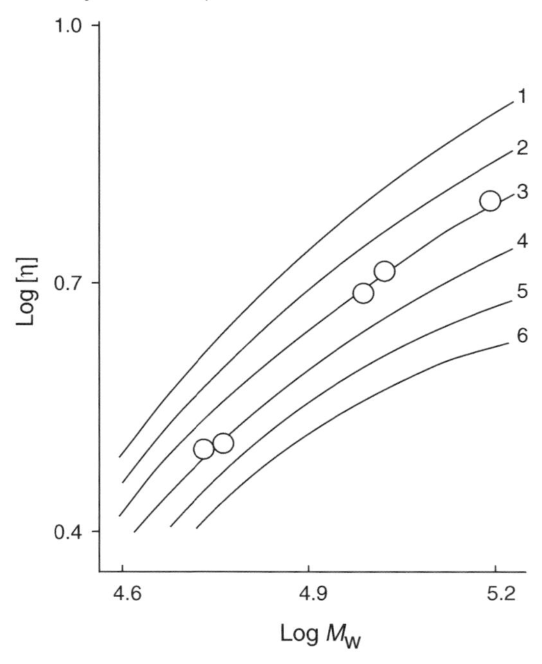

Figure 2.3 Comparison between experimental and calculated values of the intrinsic viscosity of cellulose in DMAA–LiCl mixture. The curves correspond to the theoretical dependence of log[η] on M_W for worm-like cylinders with $d=5.8$ A, $M_L=162$ and the magnitude of the persistent length of (1) 140, (2) 120, (3) 110, (4) 100, (5) 90 and (6) 80 Å. Circles denote experimental values of [η] for monodisperse samples of cellulose in DMAA–5% LiCl mixture

condition that $[\eta]^{\text{exper}} = [\eta]^{\text{theor}}$. Determination of *d* and *A* described by Papkov [32] was carried out by variation of *d* and *A* within 'reasonable' limits. Because of the ambiguity associated with the variation method, most investigators [15, 19, 35] have calculated *A* using the approximate expressions (2.11) and (2.13).

3 INFLUENCE OF THE MOLECULAR CHARACTERISTICS OF CELLULOSE AND ITS DERIVATIVES ON EQUILIBRIUM RIGIDITY IN SOLUTION

Values of diameters and lengths of Kuhn segments of macromolecules of cellulose and its derivatives from the literature [36–65] which have been calculated by methods described in Section 2.2 are given in Table 2.2. It is noteworthy that the data given for the same polymer by different authors differ appreciably. For example, the values of the parameter of equilibrium rigidity for cellulose in

Table 2.2 Length of the statistical Kuhn segment (A) and chain diameter (d) of cellulose and its derivatives

No.	Polymer: cellulose/ derivative	Molecular characteristics of polymer [a]	Solvent	Conditions	A/Å	d/Å	Reference
1	Cellulose	DS = 0	Cadoxen		76	8.0	36
2	Cellulose	DS = 0	Cadoxen		142		37
3	Cellulose	DS = 0	Cadoxen		100	16	38
4	Cellulose	DS = 0	Cadoxen		91	7	40
5	Cellulose	DS = 0	DMAA + LiCl	$c_{LiCl} = 5\%$	200		26
6	Cellulose	DS = 0	FeTNa		120	28	38
7	Cellulose	DS = 0	FeTNa		144		39
8	Cellulose	DS = 0	Cadoxen		50–150		41
9	Cellulose	DS = 0	FeTNa		120–250		41
10	Cellulose	DS = 0	DMAA + LiCl		165–250		41
11	Cellulose	DS = 0	CF_3COOH-$C_2H_4Cl_2$		170		1
12	Cellulose	DS = 0	*N*-Methylmorpholine-*N*-oxide		70–100		41
13	Cell. nitrate	DS = 2.0	Ethyl acetate		102	9.5	42
14	Cell. nitrate	DS = 2.3	Ethyl acetate		160	8	43
15	Cell. nitrate	DS = 2.4	Ethyl acetate		236	11	44
16	Cell. nitrate	DS = 2.58	Ethyl acetate	25 °C	250	17	45
17	Cell. nitrate	DS = 2.7	Ethyl acetate		254	18.6	45
18	Cell. nitrate	DS = 2.7	Ethyl acetate		200		46
19	Cell. nitrate	DS = 3.0	Ethyl acetate		236		39
20	Cell. nitrate	DS = 2.7	Butyl acetate		260		46
21	Cell. nitrate	DS = 1.9	Methyl ethyl ketone		113	10.5	47
22	Cell. nitrate	DS = 1.14	DMAA + LiCl	$c_{LiCl} = 6\%$	106	8.5	44
23	Cell. nitrate	DS = 2.55	Acetone		200	21	38
24	Cell. nitrate	DS = 2.55	Acetone		176		48
25	Cell. nitrate	DS = 2.9	Acetone		332		48
26	Cell. nitrate	DS = 2.91	Acetone		260	26	38
27	Cell. nitrate	DS = 3.0	Acetone		264		37

Table 2.2 (*continued*)

No.	Polymer: cellulose/derivative	Molecular characteristics of polymer [a]	Solvent	Conditions	A/Å	d/Å	Reference
28	Ethyl cell.	DS=2.27	Ethyl acetate		175	12	49
29	Ethyl cell.	DS=2.6	Bromoform		204		46
30	Ethyl cell.	DS=2.6	DMF		280		46
31	Ethyl cell.	DS=2.75	Acetone		168	9.5	50
32	Cyanoethyl cell.	DS=2.6	Acetone		250	8.3	51
33	Hydroxyethyl cell.		Water		182		52
34	Hydroxyethyl cell.		Water		100	29	38
35	Ethylhydroxyethyl cell.		Water		100	29	38
36	Ethylhydroxyethyl cell.		Water		180		52
37	Hydroxypropyl cell.		DMAA	25 °C	130		33
38	Hydroxypropyl cell.		DMAA	70 °C	90		33
39	Acetoxypropyl cell.		Dimethyl phthalate	25 °C	150	4	27
40	Acetoxypropyl cell.		Dimethyl phthalate	105 °C	88	10	27
41	Acetoxypropyl cell.		Dimethyl phthalate	150 °C	58	16	27
42	Cell. carbanilate	DS=2.3	Ethyl acetate		154	13	53
43	Cell. carbanilate	DS=3.0	Cyclohexanone	73 °C		77	54
44	Cell. carbanilate	DS=3.0	Cyclohexanone		160		55
45	Cell. carbanilate	DS=3.0	Dioxane		245		55
46	Cell. carbanilate	DS=3.0	Dioxane + methanol		245		56
47	Cell. carbanilate	DS=3.0	DMSO		338		57
48	Cell. carbanilate	DS=3.0	Anisole	94 °C	65		54
49	Cell. carbanilate	DS=3.0	Acetone		205		55
50	Cell. acetate		Trifluoroethanol		111		37
51	Cell. acetate		Acetone		119		37
52	Cell. acetate	DS=2.2	DMF		230		58
53	Cell. acetate	DS=2.87	Chloroform		170		58
54	Cell. acetate	DS=3.0	Acetic acid	25 °C	150		59
55	Cell. acetate	DS=3.0	Acetic acid	60 °C	90		59

56	Cell. acetate	DS = 0.49	DMAA		60	24	38
57	Cell. acetate	DS = 2.46	DMAA		100	36	38
58	Cell. acetate	DS = 2.92	DMAA		120	34	38
59	Carboxymethyl cell.	DS = 0.9	Cadoxen	26 °C	120	24	60
60	Carboxymethyl cell.	DS = 0.9	Cadoxen + water	26 °C	170	34	60
61	Cell. butyrate	DS = 2.9	Methyl ethyl ketone		200	7.0	61
62	Cell. benzoate	DS = 2.2	Dioxane		180	40	62
63	Cell. benzoate	DS = 3.0	Chloroform		216		46
64	Cell. monophenyl acetate	DS = 2.2	Benzene		110	40	62
65	Cell. dimethyl phosphonocarbamate	DS = 2.0	Water + NaCl	$c_{NaCl} = 0.2\ mol\ dm^{-3}$	165	9.0	64
66	Cell. xanthate		Water + DMSO	$c_{DMSO} = 90\%$	108		37
67	Cell. caproate		Dioxane		95		37
68	Methylol cell.		DMSO		320		64
69	Methyl cell.		Water		182		52
70	Cell. phenylcarbamate	DS = 1.9	DMF		170		46
71	Cell. monophenyl acetate	DS = 2.5	Ethyl acetate		180		46
72	NaCMC		Water + NaCl	$J \to 8\ mol\ dm^{-3}$	100	30	38
73	NaCMC	DP 120	Water + NaCl	$J = 0.1\ mol\ dm^{-3}$	254		65
74	NaCMC	DP 112	Water + NaCl	$J = 0.1\ mol\ dm^{-3}$	249		65
75	NaCMC	DP 98	Water + NaCl	$J = 0.1\ mol\ dm^{-3}$	224		65
76	NaCMC	DP 81	Water + NaCl	$J = 0.1\ mol\ dm^{-3}$	208		65
77	NaCMC	DP 55	Water + NaCl	$J = 0.1\ mol\ dm^{-3}$	165		65
78	NaCMC	DP 120	Water + NaCl	$J = 0.34\ mol\ dm^{-3}$	190		65
79	NaCMC	DP 112	Water + NaCl	$J = 0.34\ mol\ dm^{-3}$	177		65
80	NaCMC	DP 98	Water + NaCl	$J = 0.34\ mol\ dm^{-3}$	148		65
81	NaCMC	DP 81	Water + NaCl	$J = 0.34\ mol\ dm^{-3}$	132		65
82	NaCMC	DP 55	Water + NaCl	$J = 0.34\ mol\ dm^{-3}$	124		65

[a] DS = degree of substitution;
DP = degree of polymerisation;
DMAA = dimethylacetamide;
FeTNa = 'iron sodium tartrate'.

cadoxene vary from 76 to 142 Å [36–40]. Values of A for different cellulose derivatives show significant differences due to the differences in the degree of substitution and in the nature of the solvent used. It is difficult to distinguish the various effects on the parameters due to the influence of type of substituent, degree of substitution, molecular mass and solvent on the equilibrium rigidity of cellulose and its derivatives.

However, the data in Table 2.2 show that macromolecules of cellulose derivatives possess an increased equilibrium rigidity. The values of Kuhn segment lengths, A, that are characteristic of them (100–300 Å) are an order of magnitude higher that values that are typical of flexible-chain polymers. The flexibility of the molecular chain of cellulose and its derivatives is determined by the degree to which hindered rotation of pyranose cycles around two bridge bonds OC_1 and OC_4 is possible (see Figure 5.11). At the same time, the parameter of equilibrium rigidity of the cellulose derivatives exceeds considerably the Kuhn segment length for non-hindered internal rotation $A_f = 11$ Å [9]. This shows that the interaction of side groups of polyglucoside chain which prevents the rotation around valence bonds of the main chain does not contribute to the normal steric hindrance. Evidently, an important role in these interactions is played by hydrogen bonds which close a molecular chain into a cycle.

IR spectroscopic measurements indicate [66] that additional cyclization of the cellulose molecules occurring in the crystalline state is due to the formation of intramolecular hydrogen bonds (O'_3O_5) and ($O_2O'_5$) (see Figure 5.11). According to Shakhparonov *et al.* [67], the hydrogen bond (O'_3O_5) determines the conformational state of the macromolecules of cellulose acetates in solution. Some of the experimental behaviour of cellulose and its derivatives has been explained in terms of the breaking or the formation of intramolecular hydrogen bonds [36, 40]. For instance, Lyubina *et al.* [36] explained the increase in the rigidity of cellulose dissolved in an aqueous solution of cadoxene by the reduction of intramolecular hydrogen bonds which were destroyed by complexing with cadoxene.

Values of the parameter of equilibrium rigidity, A, from experimental measurements on various cellulose derivatives, are plotted against degree of substitution in Figure 2.4. This graph indicates the possibilities of variation of the flexibility of the glucopyranose chain by selecting the appropriate solvent and substituent.

It can be assumed that the introduction of substituents into a cellulose molecule creates additional steric hindrance to the mutual rotation of pyranose rings which can result in the increase in the chain equilibrium rigidity. However, if one compares the behaviour of cellulose carbanilate ($R = -CONHC_6H_5$), ethyl cellulose ($R = -C_2H_5$) and nitrate ($R = -NO_3$) with similar degrees of substitution (DS = 2.3) in ethyl acetate, it is seen that their equilibrium rigidities are within experimental error of each other (154, 175 and 160 Å, respectively [43, 49, 53]). There are data showing the increase in the rigidity of the main chain of comb-like polymers with increasing dimensions of side mesogenic groups [9]. However, it is apparently secondary cyclization of the cellulose structure by intramolecular hydrogen bonds

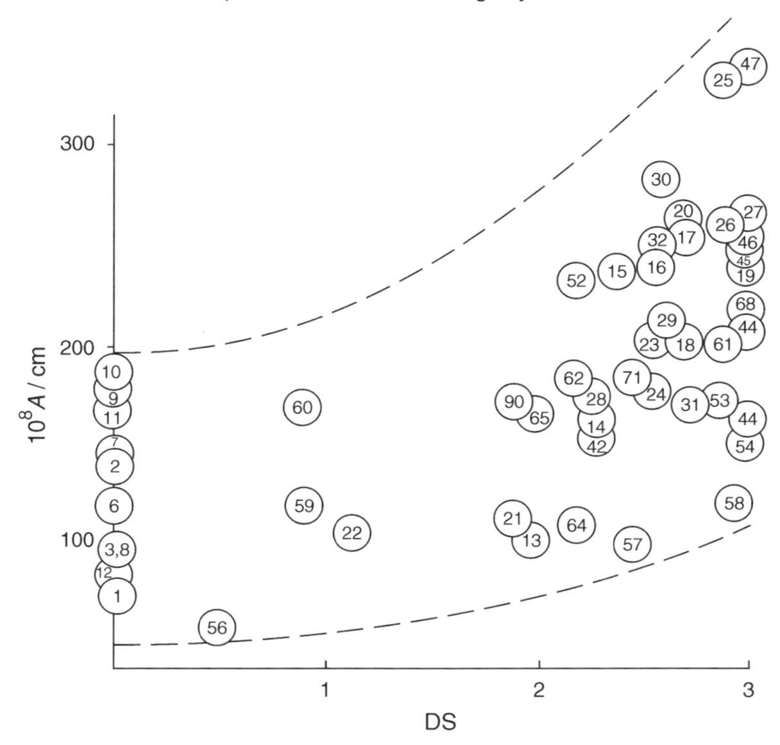

Figure 2.4 Dependence of the length of the statistical Kuhn segment on the degree of substitution of hydroxyl groups of the glucopyranose ring in cellulose derivatives. Numbering of the experimental points corresponds to that in Table 2.2

rather than steric effects due to the size of substituting groups which determine the flexibility of the chain.

Probably, introduction of substituents into the glucopyranose chain leads to a change in the localization of H-bonds within the chain: old ones are destroyed and new ones formed. Formation of new H-bonds is possible both between newly introduced groups, as in the case of the carbanilate, and also between the groups and residual hydroxyls of the pyranose cycles, as in the case of cellulose acetate. In the former case, the increase in the degree of substitution causes an increase in the chain skeleton rigidity. Burchard [11, 57] suggested that the high equilibrium rigidity of cellulose carbanilate ($A = 338$ Å in DMSO [63]) was due to secondary cyclization resulting from hydrogen bonds of the type $>NH \rightarrow O{=}C<$. In the case of cellulose acetate, an increase in the degree of substitution is accompanied by extensive change in the chain rigidity. Introduction of acetyl groups into a cellulose macromolecule leads to the increase in the rigidity of its chain due to the formation of —OH→O=C< hydrogen bonds. The number of newly formed hydrogen bonds between acetyl and hydroxyl groups is maximum at DS < 3. This corresponds to the highest value of the chain equilibrium rigidity. The magnitude of DS depends to a large extent on the evenness of distribution of the substituents within the chain. According to the data of Shakhparonov *et al.* [67] for cellulose acetates in dichloromethane, the maximum rigidity is achieved at DS = 2.90. They found that at DS $= 2.70[\eta] = 285\ \text{cm}^3\ \text{g}^{-1}$, at DS $= 2.90[\eta] = 350\ \text{cm}^3\ \text{g}^{-1}$ and at DS $= 3.0[\eta] = 2.75\ \text{cm}^3\ \text{g}^{-1}$. They also made a qualitative estimate of the number of intramolecular H-bonds by adding methanol to the solutions in dichloromethane. Introduction of methanol led to the rupture of ($O_3'O_5$) hydrogen bonds and caused a decrease in the dimensions of the macromolecular tangles. Cellulose triacetate solutions did not behave in this way. The authors stated that this was evidence for the absence of cyclization of molecules of cellulose triacetate by intramolecular H–bonds.

Saito [38] found that the chain rigidity of cellulose acetates in dimethylacetamide is doubled when the degree of substitution increases from 0.49 to 2.92 (Figure 2.5). Further experimental values of A are needed before there can be a complete understanding of the relationship between degree of substitution and chain rigidity of cellulose acetate.

Figure 2.5 shows the dependence of the equilibrium rigidity of molecules of cellulose nitrates in various solvents on the degree of substitution. There is more than a 200% increase in A on going from singly to doubly substituted cellulose nitrates. This strong dependence of the cellulose nitrate conformation on its degree of substitution was considered by Pearson and Moore [68], but the reason remains unknown. Understanding the pattern of influence of the solvent on the skeletal rigidity of cellulose nitrate molecules is complicated because of considerable differences in the data obtained by different authors under similar conditions. For example, according to Schulz and Penzel [48], for nitrate with DS = 2.9 in acetone, $A = 332$ Å; according to Saito [38], for nitrate with DS = 2.91 in acetone, $A = 260$ Å. Nevertheless, it can be considered that the rigidity of the cellulose nitrate chain, on average, increases significantly in the following series of solvents: acetone < ethyl acetate < methyl ethyl ketone < dimethylacetamide + 6% LiCl (Figure 2.5).

The influence of the solvent on the macrochain conformation was analysed in greater detail by Kamide and Saito [69]. They showed that there was a considerable increase in the rigidity of the cellulose diacetate (DS = 2.46) chain the more polar is the solvent (Figure 2.5). There is also a decrease in the equilibrium rigidity of the polyelectrolytes, sodium salts of carboxymethylcellulose (NaCMC), with increasing ionic strength of the solvent [38, 65]. For solutions of NaCMC in aqueous solutions of sodium chloride $A = 254$ Å at $I = 0.1\ \text{mol}\,\text{l}^{-1}$, $A = 190$ Å at $I = 0.34\ \text{mol}\,\text{l}^{-1}$ and $A = 100$ Å at $I \rightarrow \infty$. If one takes into account that the increase in the ionic strength of an aqueous solution of sodium chloride leads to a decrease in its dielectric permittivity, then the data obtained elsewhere [65] agree qualitatively with Kamide and Saito's measurements.

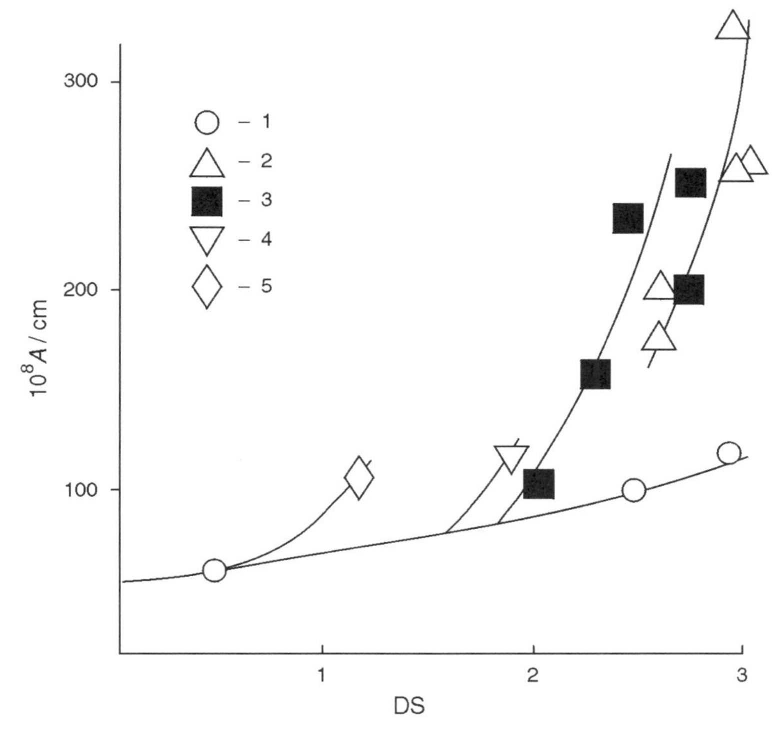

Figure 2.5 Dependence of the equilibrium rigidity of the glucopyranose chain on the degree of substitution, DS. (1) Cellulose acetates in DMAA; (2) cellulose nitrates (CN) in acetone; (3) CN in ethyl acetate; (4) CN in methyl ethyl ketone; (5) CN in DMAA–6% LiCl mixture [78]

4 REGULARITIES IN THE INFLUENCE OF MIXED SOLVENTS ON THE CONFORMATIONAL PARAMETERS AND PHASE TRANSITIONS OF CELLULOSE DERIVATIVES

The tendency towards lyotropic mesomorphism in polymeric systems is determined to a considerable degree by the degree of anisodiametry of macromolecules. The analysis of macromolecular form and of the influence of various physico-chemical parameters on it permits the development of the lyotropic LC state in non-aqueous solutions of cellulose and its derivatives to be controlled.

As was mentioned earlier, there are two basic methods for the calculation of the conformational parameters (persistent length q and hydrodynamic diameter d) of macromolecules, based on the investigations of their hydrodynamic behaviour in solutions. The first involves the use of an analytical expression [77] that establishes a linear relation between the intrinsic viscosity of a solution and the molecular mass of polymer: $([\eta], M) = f(M^{1/2})$. Here, the determination of q and d requires the estimation of the slope of a straight line and of the intercept on the ordinate axis. The second method [70] involves the variation of q and d in the theoretical equation of Yamakawa and Fujii for the intrinsic viscosity of worm-like cylinders. In the latter case, there is an assumption that a combination of the q and d parameters which give a value of the intrinsic viscosity $[\eta]$ equal to the experimental values does, in fact, correspond to the conformational state of the macromolecules.

Calculation by the first method requires $[\eta]$ values of several (not less than 6–8) fractions of polymer with a definite molecular mass. The method of substitutions enables one to determine the conformational parameters on the basis of the experimental values of the intrinsic viscosity of one fraction with a known molecular mass M. However, a value of d obtained by independent experimental methods, usually from measurements on the solid, is used in this method. It may not always be correct to assume that the value of d in solution is the same as the value for the condensed phase.

Myasoedova and Zaikov [71, 79] have developed a new calculation method for conformational parameters of cellulose derivatives in non-aqueous solutions which does not have the limitations mentioned above. The axial ratio of macromolecules is known to have a considerable influence on the value of the critical concentration for the formation of an LC phase in solution. At this concentration the equation for the intrinsic viscosity developed by Yamakawa and Fujii [16] [Eq. (2.23)] and the equation of Khokhlov and Semyonov quoted by Khokhlov [4] [Eq. (2.24)] must both be satisfied. These conditions make it possible to determine q and d of macromolecules from the two equations, i.e.

$$[\eta] = \Phi_{\infty}(L/A)^{3/2}A^3M^{-1} \tag{2.23}$$

$$V_2^* = \frac{d}{A}\left[\frac{3.34 + 11.3(L/A) + 4.06(L/A)^2}{(L/A)(1 + 0.387L/A)}\right] \tag{2.24}$$

where Φ is the viscosimetric constant, L is the contour length of a chain, $A = 2q$ is the length of the Kuhn segment and V_2^* is the critical mole fraction concentration of the semi-rigid polymer in solution.

The calculation of d and q for cellulose triacetate (CTA) macromolecules in solutions of different composition will be given as an example of this method. Values of critical concentrations and intrinsic viscosities of dilute CTA solutions in trifluoroacetic acid and dichloromethane have been published [79]. Two assumptions were made:

- approximate equality between the molecular masses of CTA samples taken initially and in solutions;
- the length of the projection of a monomeric link on to the direction of glycopyranosic chain is assumed to equal 5.15 Å [72].

Table 2.3 Values of persistent length q and hydrodynamic diameter d of CTA macromolecules in mixed solvents $CF_3COOH-CH_2Cl_2$ at 283–303 K

	$x_{CH_2Cl_2} = 0.00$		$x_{CH_2Cl_2} = 0.13$		$x_{CH_2Cl_2} = 0.25$	
T/K	q/Å	d/Å	q/Å	d/Å	q/Å	d/Å
283	58	2.4	55	2.1	74	2.7
288	57	2.4	49	2.0	73	2.7
293	54	2.4	44	1.8	72	2.8
298	54	2.4	39	1.6	68	2.7
303	51	2.3	34	1.5	65	2.6
313	48	2.3				
318	47	2.2				
	$x_{CH_2Cl_2} = 0.36$		$x_{CH_2Cl_2} = 0.47$		$x_{CH_2Cl_2} = 0.67$	
T/K	q/Å	d/Å	q/Å	d/Å	q/Å	d/Å
283	83	2.0	71	2.5	77	3.0
288	78	2.2	68	2.5	76	3.4
293	76	2.4	66	2.5	76	3.7
298	72	2.6	64	2.5	72	4.0
303	70	2.8	62	2.5		

In conformity with the above assumptions, $M_W = 1.84 \times 10^5$ and $L = 3.35 \times 10^3$ Å. The values of q and d obtained from the calculations are given in Table 2.3 [79].

The calculated values of the parameter of equilibrium rigidity of CTA macromolecules agree qualitatively with the literature data. According to Saito [38], cellulose acetate with DS = 2.92 in DMAA is characterized by $A = 120$ Å. According to the data presented here, a Kuhn segment of CTA at 298 K has a value of A from 108 to 144 Å, depending on the mixed solvent composition.

It is of interest that an increase in the proportion of dichloromethane in the mixture with trifluoroacetic acid leads to increasing equilibrium rigidity of CTA macromolecules. In the case of the solvent mixtures studied, the equilibrium rigidity of CTA molecules decreases with increasing temperature. The effect of increase in temperature on the hydrodynamic diameter of CTA in $CF_3COOH-CH_2Cl_2$ mixtures with various compositions varies. An increase in temperature has practically no effect on the value of d in mixtures with dichloromethane content below 0.4–0.5 mole fraction. In the systems with higher content of dichloromethane, where the persistent length of CTA molecules exceeds 70 Å, the hydrodynamic diameter increases with increase in temperature (Figure 2.6).

The mixed solvent with dichloromethane content above 0.4–0.5 mole fraction is interesting not only from the point of view of the anomalous dependence of d on T, but also for the peculiarities of the phase transitions of the cellulose triacetate. An

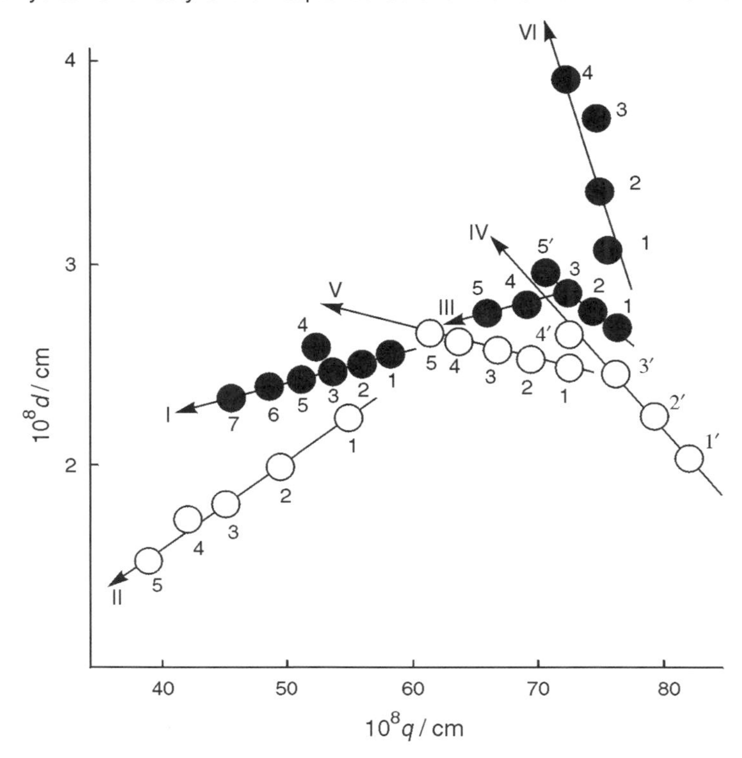

Figure 2.6 Diagram showing axial ratios of CTA macromolecules in mixed solvent $CF_3COOH-CH_2Cl_2$ with compositions (I) 0.00, (II) 0.13; (III) 0.25, (IV) 0.36, (V) 0.47 and (VI) 0.67 mole fraction of CH_2Cl_2 at temperatures (1) 283, (2) 288, (3) 293, (4) 298, (5) 303, (6) 313 and (7) 318 K

investigation of the rheological behaviour of the system CTA–$CF_3COOH-CH_2Cl_2$ was carried out by Bueche [73]. The method of interpreting rheological data developed by Saito [38] was used to estimate the phase state of the system under conditions of flow. In this interpretation it is assumed that the temperature at which minimum viscosity is found is the temperature at which an anisotropic phase appears. The temperature at which maximum viscosity occurs is taken to be the temperature of disappearance of an isotropic phase. Application of rheological criteria for the estimation of isotropic–anisotropic phase transitions enables phase diagrams of multicomponent systems of cellulose derivatives to be obtained under conditions of shear deformation. Such rheological criteria include the existence of the fluidity limit in the curves of flow of anisotropic solutions and the extreme character of the temperature and concentration dependences of viscosity.

When dissolved in pure trifluoroacetic acid, the transition of CTA macromolecules into a highly oriented state occurs over a narrow temperature and polymer concentration range. The region of co-existence of the isotropic and anisotropic phases is limited (see Chapter 1). For a particular mixed solvent, the concentration at which the phase transition occurs is proportional to the temperature. One maximum is observed in the concentration and temperature dependences of viscosity of CTA solutions in CF_3COOH. When the mole fraction of CH_2Cl_2 is increased to above 0.5 in mixtures with CF_3COOH, two maxima appear in the concentration dependences, while there is only one maximum in the temperature dependences. As the temperature increases, the first maximum in the concentration–viscosity curve degenerates into a shoulder and the second is displaced to higher polymer concentrations. The complicated character of the concentration dependence of viscosity may indicate a consecutive series of phase transitions over the temperature interval between the ternary (T_t) and critical (T_c) points. Calculations [4] for persistent molecules with axial ratio $A/d < 50$ support this hypothesis. Calculations also support the hypothesis that isotropic liquid phase stratification occurs between T_t and T_c before liquid crystalline, supramolecular ordering of the macromolecules.

5 INFLUENCE OF DIELECTRIC PROPERTIES OF SOLVENTS ON THE EQUILIBRIUM RIGIDITY OF MOLECULES OF CELLULOSE DERIVATIVES IN SOLUTION

The ability to select solvents in which macromolecules are likely to adopt the lyotropic LC state could have important practical advantages. Estimation of the influence of nature of the solvent on the equilibrium rigidity of macromolecules is one step towards this goal. The behaviour of hydroxypropylcellulose (HOPC) in non-aqueous solutions is an example. Experimental measurements have been made of the viscosity of dilute solutions of HOPC ($M_w = 120\,000$, degree of substitution 3) in water, ethanol, dimethylacetamide (DMAA) and dichloromethane [74].

The dependence of the reduced viscosity on HOPC concentration in the solvents mentioned is non-linear. The deviation from the linear dependence is predicted by Martin's equation:

$$\frac{\eta_{spec}}{C} = [\eta]e^{K_M[\eta]C} \tag{2.25}$$

This relationship is observed to a greater degree for HOPC solutions in dichloromethane and DMAA and to a smaller degree for those in water and ethanol. Owing to the non-linear character of $\eta_{red/C} = f(C)$ in the region of low HOPC concentrations, values of intrinsic viscosities were calculated by extrapolation to infinite dilution of reduced viscosity for concentrations of 4×10^{-3} to $1 \times 10^{-2}\,g\,cm^{-3}$. The calculated values of $[\eta]$ and K_M are given in Table 2.4 [74].

Since HOPC is a semi-rigid-chain polymer, the calculation of the segment rigidity of polyglucosidic chain was carried out by use of the Yamakawa–Fujii

Table 2.4 Values of the intrinsic viscosity and Martin's constants, K_M, for HOPC solutions in various solvents

T/K	Water		Ethanol		DMAA		Dichloromethane	
	$[\eta]$/ $cm^3\ g^{-1}$	K_M	$[\eta]$/ $cm^3\ g^{-1}$	K_M	$[\eta]$/ $cm^3\ g^{-1}$	K_M	$[\eta]$/ $cm^3\ g^{-1}$	K_M
288	219	0.28	334	0.10	288	0.21	358	−0.03
293	204	0.35	302	0.16	235	0.16	369	−0.03
298	187	0.42	330	0.18	—	—	284	−0.03
303	158	0.52	308	0.15	221	0.19		
313	113	0.94			198	0.22		
318					167	0.35		

Table 2.5 Values of Kuhn segment for HOPC molecules in various solvents [74]

T/K	Kuhn segment/Å			
	Water	Ethanol	DMAA	Dichloromethane
288	210	370	220	410
293	190	320	230	430
298	170	370	—	460
303	160	330	240	250
308	140	330	210	
313	100	130	190	
318	90		150	

equation. The value of the HOPC chain diameter, d, was taken to be 12.8×10^{-8} cm [52]. Calculated values of a Kuhn segment for HOPC molecules are given in Table 2.5.

Figure 2.7 shows the dependence of the size of the Kuhn segment of HOPC on the dielectric permeability of the solvent. The tendency for a decrease in the segment rigidity of HOPC chain with increasing polarity of solvent is obvious. This tendency is in good agreement with the proportional dependence of A on ε^{-1}, that was found for polyisocyanate [75], and also with the results obtained for solutions of hydroxypropylcellulose acetate [76].

However, the data of Kamide and Saito [69] for cellulose diacetate (degree of substitution 2.46) shows an increase in the equilibrium rigidity of the polymer chain on going from tetrahydrofuran to the more polar acetone and DMAA. Values of A for CA and for HOPC show different dependences on ε. It follows that the type of group that is substituted into the cellulose chain must also be taken into account when attempting to predict the conditions for the formation of anisotropic phases.

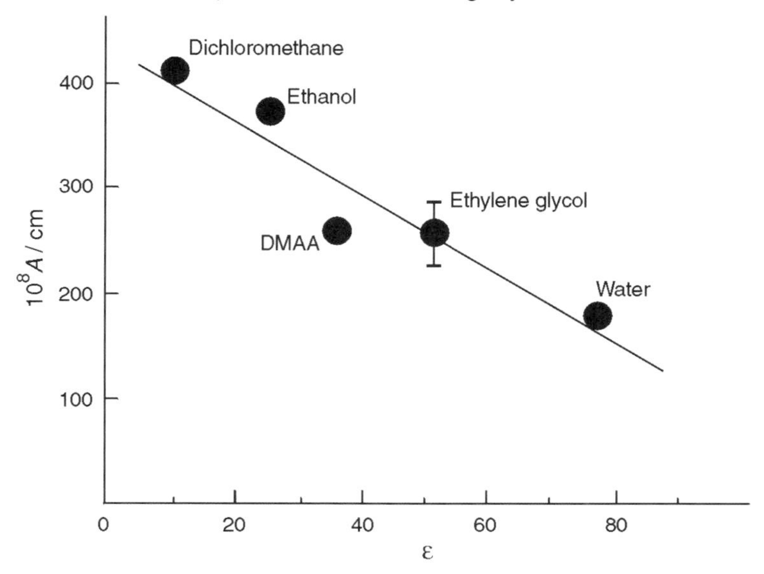

Figure 2.7 Dependence of the size of the Kuhn segment of HOPC on the dielectric permeability of the solvent

REFERENCES

1. Krestov G. A., Timofeyeva G. I., Myasoedova V. V., Alexeyeva O. V., Osipov S. A., Ermakova V. D., Fronchek E. V. *Dokl. Akad. Nauk SSSR* **1989**, *305*, 131–34.
2. Gotlib Yu. Ya., Darinskii A. A., Svetlov Yu. E. *Physical Kinetics of Macromolecules*, Khimiya, Leningrad **1986**.
3. Smirnova V. N., Iovleva M. M., Kopyev M. A., Figner G. G., Baksheyev I. P. *Khim. Volokna* **1988**, *3*, 23–24.
4. Khokhlov A. R. *Statistical Physics of Macromolecules* MGU, Moscow **1985**.
5. Kuhn W. *Kolloid-Z.* **1934**, *68*, 2–15.
6. Guth E., Mark H. *Monatsh. Chem.* **1934**, *65*, 93–105.
7. Marikhin V. A., Myasnikova L. P. *Super-Molecular Structure of Polymers* Khimiya, Leningrad **1977**.
8. Eyring H. *Phys. Rev.* **1932,** *39*, 746–48.
9. Tsvetkov V. N. *Rigid-Chain Polymer Molecules* Nauka, Leningrad **1986**.
10. Tsvetkov V. N., Shtennikova I. N., Skazka V. S., Ryumtsev E. I. *Polym. Sci.* **1968**, *16*, 3205–17.
11. Burchard W. *Makromol. Chem.* **1965**, *88*, 11–28.
12. Dotu P. *Proc. Natl. Acad. Sci. USA* **1956**, *42*, 791–800.
13. Kratky O., Porod G. *Recl. Trav. Chim.* **1949**, *68*, 1106–22.
14. Porod G. *Monatsh. Chem.* **1949**, *2*, 251–58.
15. Novakovskaya V. B., Strelina I. A. *Vysokomol. Soyedin.* **1988**, *30*, 2467–72.
16. Yamakawa H., Fujii M. *Macromolecules* **1974**, *7*, 128–35.

17. Rudkovskaya G. D., Shabsels B. M., Baranovskaya I. A., Ulyanova N. N., Lyubina S. Ya., Bezrukova M. A., Vlasov G. P., Eskin V. E. *Vysokomol. Soyedin.* **1989**, *31*, 133–39.
18. Tsvetkov V. N., Andreyeva L. N., Lavrenko P. N., Okatova O. V., Belyayeva E. V., Bilibin A. Yu., Skorokhodov S. S. *Vysokomol. Soyedin.* **1988**, *30*, 1263–68.
19. Pogodina N. V., Evlampiyeva N. P., Tsvetkov V. N., Korshak V. V., Vinogradova S. V., Rusanov A. L., Ponomaryov I. I., Lekae T. V. *Vysokomol. Soyedin.* **1988**, *30*, 1198–1205.
20. Tsvetkov V. N., Andreyeva L. N., Bushin S. V., Belyayeva E. V., Cherkasov V. A., Mashonin A. I., Bilibin A. Yu., Skorokhodov S. S. *Vysokomol. Soyedin.* **1988**, *30*, 713–21.
21. Yamakawa H., Yoshizaki T. *Macromolecules* **1980**, *13*, 633–43.
22. M. Bickles, L. Segal (eds) *Cellulose and Its Derivatives* Mir, Moscow, **1974**.
23. Stockmayer W. H., Fixman M. *J. Polym. Sci., Part C* **1963**, *1*, 197–204.
24. Tsvetkov V. N., Lezov A. V. *Vysokomol. Soyedin. Krat. Soobshch.* **1984**, *26*, 494–98.
25. Bushin S. V., Tsvetkov V. N., Lysenko E. B., Emelyanov V. N. *Vysokomol. Soyedin.* **1981**, *23*, 2494–2503.
26. Jerbojevich M., Casani A., Conio J., Ciferri A., Bianchi E. *Macromolecules* **1985**, *18*, 646–50.
27. Laivins G. V., Gray D. G. *Macromolecules* **1985**, *18*, 1746–52.
28. Bohdanecky M., Kovar T. *Macromolecules* **1983**, *16*, 1483–92.
29. Alexeyeva O. V., Timofeyeva G. I., Osipov S. A., Myasoedova V. V. in *Abstracts of All-Union Conference Chemistry and Application of Non-Aqueous Solutions, Ivanovo* **1986**, *3*, 437.
30. Myasoedova V. V., Alexeyeva O. V., Krestov G. A. in *V All-Union Meeting on Chemistry of Non-Aqueous Solutions of Inorganic and Complex Compounds, Rostov-on-Don* **1985**, 226–27.
31. Klenkova N. I., Bludova O. S., Matveeva N. A. *Cellul. Chem. Technol.* **1982**, *16*, 615–30.
32. Papkov S. P. *Vysokomol. Soyedin. Krat. Soobshch.* **1982**, *24*, 869–73.
33. Conio J., Bianchi E., Ciferri A., Aden M. A. *Macromolecules* **1983**, *16*, 1264–70.
34. Tsvetkov V. N., Filippov A. P. *Vysokomol. Soyedin.* **1989**, *31*, 2249–55.
35. Kuznetsova G. B., Silinskaya I. G., Kallistov O. V., Kalashnikov B. O., Shirokova L. G., Efros L. S. *Vysokomol. Soyedin.* **1988**, *30*, 586–90.
36. Lyubina S. Ya., Klenin S. I., Strelina I. A., Troitskaya A. V., Khripunov A. K., Urinov E. U. *Vysokomol. Soyedin.* **1977**, *19*, 244–49.
37. Swenson H. A., Schmitt C. A., Thompson N. S. *J. Polym. Sci., Part C* **1965**, *3*, 243–52.
38. Saito M. *Polym. J.* **1983**, *15*, 213–23.
39. Flory P. J. *Principles of Polymer Chemistry* Cornell University Press, Ithaca, NY **1965**.
40. Lyubina S. Ya. in *Investigation Methods for Cellulose* Karlivan V.P. (ed.) Zinantne, Riga **1981**, 159–65.
41. Prozorova G. E., Iovleva M. M., Dibrova A. K., Belousov Yu. Ya., Petrova L. V., Papkov S. P. *Vysokomol. Soyedin.* **1986**, *28*, 412–15.
42. Pogodina N. V., Melnikov A. B., Mikryukoya O. I., Didenko S. A., Marchenko G. N. *Vysokomol. Soyedin.* **1984**, *26*, 2515–20.
43. Pogodina R. V., Lavrenko P. N., Pozhivilko K. S., Melnikov A. B., Kolobova T. A., Marchenko G. N., Tsvetkov V. N. *Vysokomol. Soyedin.* **1982**, *24*, 332–38.
44. Bushin S. V., Lysenko E. B., Cherkasov V. A., Smirnov K. P., Didenko S. A., Marchenko G. N., Tsvetkov V. N. *Vysokomol. Soyedin.* **1983**, *25*,1899–1905.
45. Pogodina N. V., Bushin S. V., Melnikov A. B., Lysenko E. B., Kolobova T. A., Marchenko G. N., Shipina O. T., Tsvetkov V. N. *Vysokomol. Soyedin.* **1987**, *29*, 299–305.

46. Shtennikova I. N. in *Investigation Methods for Cellulose* Karlivan V. P. (ed.) Zinantne, Riga **1981**, 149–58.
47. Pogodina N. V., Pozhivilko K. S., Evlampiyeva N. P., Melnikov A. B., Bushin S. B., Didenko S. A., Marchenko G. N., Tsvetkov V. N. *Vysokomol. Soyedin.* **1981**, *23*, 1252–60.
48. Schulz G. V., Penzel E. *Makromol. Chem.* **1968**, *112*, S260–S280.
49. Ryumtsev E. I., Aliyev F. M., Vitovskaya M. G., Urinov E. U., Tsvetkov V. N. *Vysokomol. Soyedin.* **1975**, *17*, 2676–81.
50. Meyerhoff G., Sutterlin N. *Makromol. Chem.* **1965**, *87*, 258–70.
51. Tsvetkov V. N., Lavrenko P. N., Andreeva L. N., Mashoshin A. I., Okatova O. V., Mikriukova O. I., Kutsenko L. I. *Eurp. Polym. J.* **1984**, *20*, 823–29.
52. Werbowyj R. S., Gray D. G. *Macromolecules* **1980**, *13*, 69–73.
53. Andreyeva L. N., Lavrenko P. N., Urinov E. U., Kutsenko L. I., Tsvetkov V. N. *Vysokomol. Soyedin. Krat. Soobshch.* **1975**, *17*, 326–30.
54. Shanbhag V. P., Ohman J. *Ark. Kemi* **1968**, *29*, 163–78.
55. Shanbhag V. P. *Ark. Kemi* **1968**, *29*, 1–22.
56. Burchard W., Husemann E. *Makromol. Chem.* **1961**, *44/46*, 358–87.
57. Burchard W. *Br. Polym. J.* **1971**, *3*, 214–20.
58. Musayev Kh. N., Akbarov Kh. I., Nikonovich G. V., Tashmukhamedov S. A. *Vysokomol. Soyedin.* **1988**, *30*, 1370–73.
59. Ryskina I. I., Vakulenko N. A. *Vysokomol. Soyedin.* **1987,** *29*, 306–31.
60. Filipp B. Lavrenko P. N., Akatova O. V., Filippova T. V., Mikryukova O. I., Petrov V. R., Dautzenberg H., Shtennikova I. N., Tsvetkov V. N. *Vysokomol. Soyedin.* **1987**, *29*, 32–38.
61. Lyubina S. Ya., Klenin S. I., Strelina I. A., Troitskaya A. V., Kurlyankina V. 1., Tsvetkov V. N. *Vysokomol. Soedin.* **1973**, *15*, 691–98.
62. Korneyeva E. V., Lavrenko P. N., Urinov E. U., Khripunov A. K., Kutsenko L. I., Tsvetkov V. N. *Vysokomol. Soedin.* **1979**, *21*, 1547–53.
63. Zakharova E. N., Kutsenko L. I., Tsvetkov V. N., Skazka V. S., Tarasova G. V., Yamshchikov V. M. *Vestn. LGU Ser. Fiz. Khim.* **1970**, *16*, 55–63.
64. Tieda D. A., Stratton R. A. in *Abstracts of the ACS Meeting of the Cellulose, Paper and Textile Chemistry Division* **1977**, 27–31.
65. Ozolinsh R. E., Karkla M. A. *Khim. Dreves.* **1984**, *3*, 68–70.
66. Marrinan H. J., Mann J. *J. Appl. Chem.* **1954**, *4*, 204–21.
67. Shakhparonov M. I., Zakurdayeva N. P., Podgorodetskii Yu. K. *Vysokomol. Soyedin.* **1967**, *19*, 1212–20.
68. Pearson G. P., Moore W. R. *Polymer* **1960**, *1*, 144–48.
69. Kamide K., Saito M. *Polym. J.* **1982**, *14*, 517–26.
70. Birshtein T. M., Zhulina E. V. *Polymer* **1984**, *25*, 1453–57.
71. Myasoedova V. V. *Int. J. Polym. Mater.* **1995**, *38*, 23–42.
72. Borisov O. V., Birshtein T. M., Zhulina E. B. *Vysokomol. Soyedin.* **1987**, *29*, 1413–18.
73. Bueche F. *J. Chem. Phys.* **1954**, *22*, 1570–76.
74. Williams M. C. *AIChE J.* **1966**, *12,* 1064–70.
75. Golovko L. I., Rumyantsev L. Yu., Shilov V. V., Kovernik G. P. *Vysokomol. Soyedin.* **1988**, *30*, 2572–77.
76. Chang D. K. *Rheology in Processing of Polymers* Khimiya, Moscow **1979**.
77. Ciferri A., Ward I. (eds) *Super-Highly-Modular Polymers* Khimiya, Leningrad **1983**.
78. Bosch T.A., Maissa P., Sizou P. *Nuovo Cim. D* **1984**, *3*, 95–103.
79. Myasoedova V. V., Zaikov G. E. in *New Approaches to Polymer Materials* Zaikov G.E (ed.) Nova Science Publishers, New York **1996**, 125–151.

CHAPTER 3

Thermochemistry of Dissolution of Cellulose in Non-aqueous Solvents

1 ENTHALPIC CHARACTERISTICS OF SOLVATION OF NON-ELECTROLYTES

In general, solvation should be considered as the sum of energetic and structural changes occurring in a system during transition into the liquid phase and resulting in the formation of a solution with a definite chemical structure [7]. Solvation of electrolytes has been studied in the greatest detail [8]. There are numerous theoretical methods of finding the standard thermodynamic characteristics of dissolution and solvation on the basis of experimental data [9–11]. Thermochemistry of solvation of non-electrolytes has been studied less extensively, with little work on polymer non-electrolytes.

The solvation enthalpy of a solute is the transfer enthalpy of a substance from the gaseous phase into the solvent and can be expressed by the following equation:

$$\Delta H_{\mathrm{solv}} = \Delta H_{\mathrm{soln}} - \Delta H_{\mathrm{evap}} \tag{3.1}$$

where ΔH_{evap} is the evaporation (sublimation) enthalpy of a substance at a given temperature, ΔH_{soln} is the dissolution enthalpy of a substance in particular solvent and ΔH_{solv} is the enthalpy of solvation of a solute by the solvent molecules.

The solvation enthalpy is an energetic characteristic of the intermolecular interaction in solution both between solute and solvent molecules and between the molecules of solvent. There are several methods of determining solvation enthalpy. One way is to represent the solvation enthalpy [12–14] as the sum of the enthalpies of cavity formation (ΔH_{cav}) and of solvent–solute interaction ($\Delta H_{\mathrm{interac}}$):

$$\Delta H_{\mathrm{solv}} = \Delta H_{\mathrm{cav}} + \Delta H_{\mathrm{interac}} \tag{3.2}$$

Another method involves the representation of ΔH_{solv} as the sum of the enthalpies of non-specific solvation (ΔH^{non-sp}_{solv}) and of specific interaction ($\Delta H^{sp}_{interac}$) [15, 16]:

$$\Delta H_{solv} = \Delta H^{non-sp}_{solv} + \Delta H^{sp}_{interac} \tag{3.3}$$

It is apparent from Eqs (3.2) and (3.3) that the enthalpy of non-specific solvation includes the enthalpy of cavity formation in the solvent. It also includes the enthalpy of solute–solvent interaction which is determined by non-specific interactions due to van der Waals forces. The enthalpy of specific interaction of solute with solvent is determined by the donor–acceptor character of bonds.

The solvating power of a solvent may be characterized by the transfer enthalpy (ΔH_{tr}) from a standard solvent into the solvent under test [5, 17, 18]:

$$\Delta H_{tr} = \Delta H_{S1-S2} = \Delta H_{solv\,S2} - \Delta H_{solv\,S1} = \Delta H_{soln\,S2} - \Delta H_{soln} \tag{3.4}$$

where S_1 and S_2 are the standard solvent and solvent under test, respectively.

Several methods for the determination of the enthalpies of specific interactions have been developed and tested experimentally. Essentially all these methods involve obtaining the enthalpic contribution from non-specific solvation followed by determination of the enthalpy of specific interaction. The latter is the difference between the enthalpy of solvation and that of non-specific interaction. This is well described in publications by Gutmann [19], Arnett and co-workers [20, 21], Duer and Beotrand [22] and Solomonov *et al.* [23].

Gutmann [19] used experimentally determined enthalpies of interaction of compounds with $SbCl_5$ in a medium of 1,2-dichloroethane to characterize the donating abilities of substances. Arnett and co-workers [20, 21] suggested that the contribution from the enthalpies of formation of hydrogen bonds between a solvent and solvent could be determined from the relationship below. They called the method the 'pure-base' method.

$$-\Delta H_{H-bond} = (\Delta H^{A}_{soln} - \Delta H^{M}_{soln})_{S_2} - (\Delta H^{A}_{soln} - \Delta H^{M}_{soln})_{S_1} \tag{3.5}$$

where A and M refer to the solute under test and the model solute, respectively. When the model compound is selected, the following must be assumed:

$$(\Delta H^{non-spA}_{solv} - \Delta H^{M}_{solv})_{S_2} = (\Delta H^{A}_{solv} - \Delta H^{M}_{solv})_{S_1} \tag{3.6}$$

As an alternative, it was proposed by Levina *et al.* [24] that the following should hold:

$$\left(\frac{\Delta H^{non-spA}_{solv}}{\Delta H^{M}_{solv}}\right)_{S_2} = \left(\frac{\Delta H^{A}_{solv}}{\Delta H^{M}_{solv}}\right)_{S_1} \tag{3.7}$$

Arnett and co-workers proposed the use of substances in which active hydrogen is substituted by a methyl group as model compounds and carbon tetrachloride as the standard solvent. Neither of these could take part in specific interaction due to hydrogen bonding. However it has been reported [22, 23] that there are difficulties in selecting a suitable model compound for investigating non-specific solvation in a range of solvents.

The non-hydrogen-bonding baseline (NHBB) method has been described by Levina *et al.* [24] and by Stephenson and Fuchs [25]. Unlike the methods of 'solvation enthalpies' and of 'pure base,' there is compensation for the dispersion and for polar interactions in this method.

The solvating power of solvents can be characterized by a series of parameters [26, 27]. The empirical scales of solvent 'force' given in these papers were found from the interactions of model substances (indicators) with solvents. These interactions were calculated from certain physical properties which included IR and UV spectra, chemical shifts in NMR spectra and interaction enthalpies. Parameters which can be used to interpret solvation enthalpies include empirical parameters of polarity, donor (DN) and acceptor (AN) numbers, molar refraction (MR), cohesion energy densities (δ) and Kamlet–Taft parameters (Π). There are interrelations between the parameters [6, 26, 27] in the case of groups of related solvents.

A method of estimating specific interactions (hydrogen bond formation) by means of correlation of the non-specific solvation enthalpy with the molar refraction (MR) has been proposed by Solomonov and co-workers [16, 23]. Hydrogen bond formation enthalpies of a series of solutes in a solvent under test is assumed to be equal to the deviation of the solvation enthalpies from a linear variation with molar refractions of the solutes in the standard solvent (CCl_4). The authors found four patterns of behaviour of organic compounds. The solvation enthalpies of various compounds dissolved in alkanes and those of alkanes of the same structural type dissolved in various solvents show a linear dependence on their molar refraction. The dipole moment of the solute does not influence the enthalpy of solvation appreciably. Dispersion interactions between solvent and solute per repeating unit of polymer depend on the structure of solute molecules and on solvent properties. The variation of the non-specific solvation enthalpy of solutes with their molar refractions in various solvents, with the exception of alkanes, is qualitatively similar to the variation of the solvation enthalpy of these compounds with their molar refractions in carbon tetrachloride.

The methods considered above give the enthalpies characteristic of the solvation processes ($\Delta H_{\text{solv}}^{\text{non-sp}}, \Delta H_{\text{interac}}^{\text{sp}}$) and sometimes can give the enthalpy of hydrogen bond formation.

Thermochemical data [23] agree well with the results of other methods of investigating solvation processes. The enthalpic characteristics of solvation of compounds in various solvents can be used to estimate the energetics of inter-particle interactions in solution. However, there are difficulties. Since donor–acceptor compounds often have high polarity, it is impossible to select a model for revealing the contribution from polar interactions. Gryzlov and Goldshtein's investigations show that in the general case, the interaction enthalpy depends on the dipole moment of the substance dissolved [28].

The complicated structure of high-molecular-mass compounds and the limited number of solvents for them makes the determination of specific solvation enthalpies of cellulose and other polysaccharides difficult.

2 MATHEMATICAL MODELS FOR CALCULATION OF THE EXCESS ENTHALPY OF MIXING

There has been much recent interest in the thermochemistry of binary mixtures [29–31] and, in particular, their application as solvents for cellulose and its derivatives [32–35]. There has been special interest in enthalpies of mixing of liquids. Extended reference data for thousands of binary mixtures have been published [36–39]. In Russia, systematic work has been carried out by Belousov and co-workers [4, 36, 39].

The enthalpy of mixing can be a source of reliable information on intermolecular interactions in solutions [40–42]. Measurements can be made with high precision, allowing reliable comparisons between experimental measurements and theoretical calculations based on molecular statistics.

The problems of interpolation and extrapolation of thermochemical data by graphical methods and of numerical methods have been considered by various authors [40–42]. An estimation of the excess enthalpy of the glass-like state of cellulose in solution to be 12.6 kJ mol^{-1} was made by Tsvetkov [42]. This is based on enthalpies of mixing of cellulose, starch and dextran with water over a wide concentration range with appropriate graphical extrapolation.

A series of equations have been proposed [43, 44] for the empirical approximation of data for enthalpies of mixing. Correlations of the concentration dependence of the excess enthalpies in binary systems have been published by Golova *et al.* [33]. In many cases a Reidlich–Kister polynomial gives good results:

$$H^{\mathrm{e}} = RTx_1x_2 \sum_{i=0}^{p-1} A_i(x_1 - x_2)^i \tag{3.8}$$

where x_1 and x_2 are the mole fractions of the components of a binary mixture and A_i is a constant of the equation. However, Eq. (3.8) is not applicable to alcohol–water mixtures and other systems with associated hydroxyl-containing components. It describes rather well the sigmoid dependence of the excess enthalpies of mixing on concentrations but cannot be used for the calculation of asymmetric concentration dependence of H^{e} on concentrations.

A function was proposed by Brandreth *et al.* [43a]:

$$H^{\mathrm{e}} = RTx_1x_2 \sum_{i=0}^{p-1} A_i x^{i/2} \tag{3.9}$$

This function was successfully applied to cyclic alkanes and alcohols but has similar faults to the Reidlich–Kister polynomial.

The Mrazek and van Ness equation [Eq. (3.10)], with more than two constants, suffers from the disadvantage of giving indefinite values of H^{e} when $x_1 \to 0$ [43].

$$H^{\mathrm{e}} = RTx_1x_2 \sum_{i=0}^{p-1} A_i x^{-i} \tag{3.10}$$

In addition, this correlation does not reproduce the sigmoid dependence of H^{e} on x found, for example, with alcohol–hydrocarbon systems.

The sum of symmetrical function (SSF) equation:

$$H^{\mathrm{e}} = RTx_1x_2 \sum_{i=1}^{p/2} A_i(x_1/a_i + a_ix_2)^{-2} \tag{3.11}$$

describes well an asymmetric dependence of H^{e} on x but is only marginally better than Eq. (3.8) for symmetric systems.

Wilson's equation [43b] is of limited application, since it is not applicable to systems with specific interactions [43].

Balk and Somsen [29] have discussed the enthalpies of solution of nine polyols in binary solvent mixtures of *N,N*-dimethylformamide and water and of one polyol in a mixture of formamide and water at 298 K over the whole mole fraction range. Deviation from linear behaviour at both high and low water contents is due to hydrophobic hydration of the apolar sites. At low water content there is an additional effect due to preferential hydrogen bonding of the hydroxy groups of the solute. The effects are described by a simple two-parameter hydrophobic hydration model and a thermodynamic theory of preferential solvation, respectively. With regard to enthalpies, the relative importance of the two solvation mechanisms appeared to depend on the co-solvent and stereochemical details of the solute.

Nikolaenko and Batalin [44] have suggested that the coefficients of Legendre polynomials should be used for the classification of the properties of liquid two component melts for storage in thermodynamic data banks.

There are various other approaches to correlate enthalpies of mixing with physico-chemical properties. These include the use of models based on lattice theory or the theory of associative equilibria [45]. Group models such as UNIFAC [46] and ASOG [47] are widely used. These can be used to calculate phase equilibria, activity coefficients and also values of the Flory–Huggins χ parameter for polymer solutions [48].

Smirnova has published detailed accounts of lattice models [45, 48]. These models are probably the most useful models for polymers solutions.

3 ENTHALPIES OF SOLUTION, DILUTION AND MIXING OF POLYSACCHARIDES: DEXTRANS IN WATER, NON-AQUEOUS AND MIXED SOLVENTS

Enthalpies of solution and of dilution yield quantitative information on the energetics of interactions between solvent molecules and macromolecules in solution. Thermal effects also give evidence for numerous conformational changes. Fredenslund *et al.* have published heats of solution and dilution of dextranes of different molecular masses in ethylene glycol, glycerine, dimethyl sulfoxide (DMSO), ethanolamine, formamide and water. Similar studies on the saccharides raffinose, maltose and glucose were also reported [47].

Integral enthalpies of solution of dextran monomers, oligomers and high polymers with molecular masses (M) of 2×10^4, 4×10^4, 7×10^4 and 11×10^4 in water, formamide (FA), DMSO and their mixtures have been measured at 298.15 K by Myasoedova *et al.* [51]. Dextran concentrations in solution were 0.05–0.3 mass%. An increase in M values of dextrans from 100 to 680 does not result in a significant change in values of ΔH_{soln}, defined as the enthalpy of solution per gram of polymer. This agrees with the theory that the degree of polymerization (DP) of the flexible-chain amorphous polymers has little influence on ΔH_{soln} when DP > 10. With decreasing DP, in the series from oligomers to monomers, there is a linear decrease in absolute values of the integral enthalpies of solution [52]. Such a dependence of ΔH_{soln} on M can be explained by a change in the conformation of dextran molecules from the statistical tangle to a worm-like one with an increase in crystallinity. As a consequence, there is an increase in an endothermal contribution due to the melting of crystalline regions. The variation of ΔH_{soln} of dextran ($M = 110\,000$) with change in concentration in various solvents has been measured by Pokrovskii [53].

The enthalpy of solution of completely amorphous dextran in a liquid can be considered to be made up of at least three contributions:

$$\Delta H_{\mathrm{soln}} = \Delta H_{\mathrm{rub}} + \Delta H_{\mathrm{conf}} + \Delta H_{\mathrm{solv}} \qquad (3.12)$$

The first contribution, ΔH_{rub}, relates to the transition of the amorphous polymer to a rubber-like state in solution. Tsvetkov *et al.* [54] estimated the value of ΔH_{rub} for amorphous dextran to be 72.1 J g^{-1} or 11.7 kJ mol^{-1}. The second contribution, ΔH_{conf}, relates to the thermal effects of the conformational transformations in solution of polysaccharide molecules. Since dextran conformers in solutions are isoenergetic [55], ΔH_{conf} can be assumed to equal zero. The third contribution, ΔH_{solv}, characterizes the solvation by solvent of hypothetically liquid dextran and depends on the intermolecular interactions between the components of the solution. The presence of numerous hydroxyl groups in the dextran molecule results in the formation of a tight solvation shell held by polymer–solvent hydrogen bonds. This specific solvation is a reason for the high negative values of ΔH_{solv} (from -54.8 J g^{-1} in FA to -82.1 J g^{-1} in DMSO).

The exothermicity of dextran solvation increases in the series ethylene glycol < glycerine < formamide ≤ water < dimethyl sulfoxide < ethanolamine. The value of ΔH_{sol} for water is close to that for formamide. The values clearly depend on the characteristics of individual solvents. It is necessary to take into account both donor and acceptor properties of liquids when interpreting polysaccharide solvation, because of the bifunctional nature of dextrane hydroxyl groups. Abakshin and Krestov [56] proposed that half the sum of DN and AN numbers should be used as the criterion of the donor–acceptor ability of a solvent. Formamide and water have values of $\frac{1}{2}(\mathrm{DN}+\mathrm{AN})$ which are fairly close, 31.9 and 36.4, respectively. In addition, formamide and water are strongly associated liquids with almost the same values (15.5 kJ mol^{-1}) for the energies of the formamide–formamide, water–water and formamide–water hydrogen bonds [14].

The increase in exothermicity of dextran solvation in DMSO, relative to that in water and formamide, is due to a complicated mutual influence of various factors. The value of Flory's interaction parameter of dextran–DMSO system is one order of magnitude higher than that for the dextran–water system [50]. The data from ^{13}C NMR relaxation measurements [57] indicate that the conformational mobility of dextran molecules in water is higher than in DMSO. The endothermal contribution to ΔH_{solv} associated with the destruction of solvent–solvent bonds in water is the least for the liquids studied.

Figure 3.1 shows the dependence on mixed solvent composition of integral solution enthalpy, ΔH°_{soln}, of dextran ($M = 110\,000$) in water–FA, water–DMSO and FA–DMSO mixtures at 298.15 K. The three systems show different behaviours. For the water–FA system, the dependence is linear. Water–DMSO and FA–DMSO systems are characterized by minima in the variation of $-\Delta H^{\circ}_{soln}$ with x_2. The

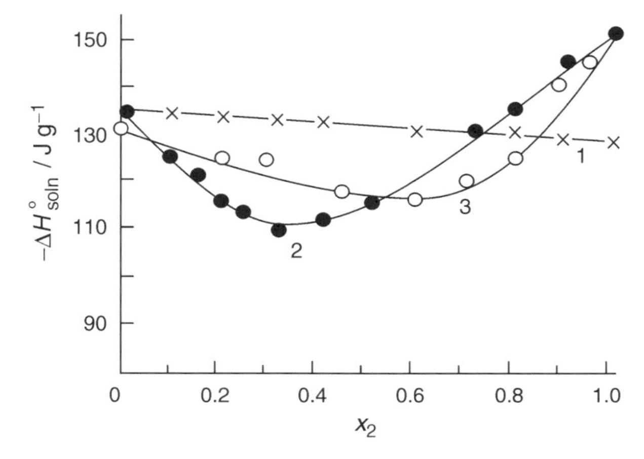

Figure 3.1 Enthalpy of solution of dextran (DP = 680) in mixtures at 298 K. 1, H_2O-FA; 2, H_2O–DMSO; 3, FA–DMSO. x_2 is the mole fraction of the second component of the mixture

different behaviour is the result of donor–acceptor interactions between components of the second two mixtures which do not occur in the water–FA mixture. The molecular interaction between components of the solvents is the strongest at the compositions at which the minima of $-\Delta H^{\circ}_{soln}$ is observed (minimum heat evolution). The enthalpies of mixing of the solvent components pass through maxima as the concentration ratios are changed. The formation of mixed associates from the solvent molecules increases as the ratio of concentrations becomes more favourable. This results in a corresponding decrease of dextran solvation and, as a consequence, in a decrease of exothermicity of solution [53].

The value of ΔH_{tr} varies linearly with mole fraction of the second solvent for the H_2O–FA system (see Figure 3.1). The values of ΔH_{tr} for the FA–DMSO and H_2O–DMMO systems pass through maxima and do not show additivity. The descending part of the curve for H_2O–DMSO system is concave. Such a variation of ΔH_{tr} with composition for solutions of low-molecular-mass saccharides and polyalcohols is, according to Balk and Somsen [58], a criterion of the existence of selective solvation. Here it means that dextran hydroxyl groups are selectively solvated by water in water–DMSO mixtures. This conclusion agrees with the findings of Basedow *et al.* [52] who proved the existence of selective solvation of dextrans in water–DMSO mixtures by gel-permeation chromatography. Figure 3.2 shows the dependence on DMSO mole fraction of the transfer enthalpies of dextran, ΔH_{tr}, from DMSO into water–DMSO and FA–DMSO mixtures.

A comparison between the variations of ΔH_{tr} with molar composition of mixed solvent indicates that selective solvation of macromolecules is absent in water–formamide but occurs in water–DMSO. This is due to the differences in the

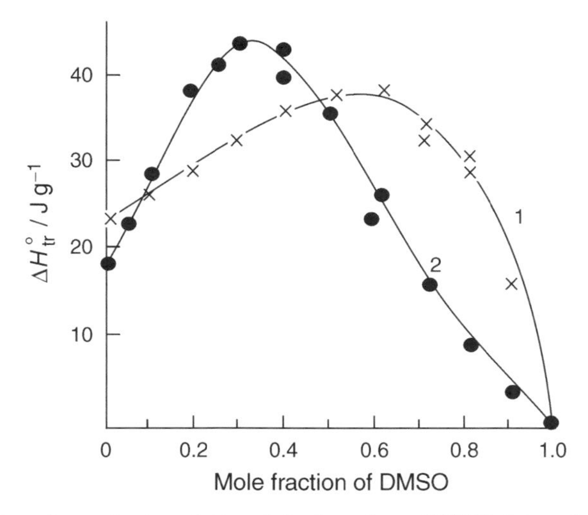

Figure 3.2 Transfer enthalpy, ΔH_{tr}, of dextran from DMSO into mixed solvents at 298 K. 1, H_2O–DMSO; 2, FA–DMSO

formation of hydrogen bonds with dextran macromolecules in the water–organic solvent and in the non-aqueous solvent mixture. In the water–DMSO system, when $x_{DMSO} > 0.5$, water–dextran hydrogen bonds prevail owing to the steric difficulties of solvation of polysaccharide by DMSO molecules.

4 INFLUENCE OF SOLVENT NATURE ON THE ENTHALPY OF INTERACTION WITH CELLULOSE AND ITS DERIVATIVES

The interaction enthalpy of cellulose with a solvent depends on the nature of electron donor–acceptor interactions in the system. This depends on both the structural peculiarities of cellulose and on the nature of the solvent. The values of the enthalpy of mixing of cellulose with various organic solvents are given in Table 3.1 [59].

The large variation in the absolute values of these enthalpy changes (Table 3.1) is due to the extent to which the crystalline regions are destroyed when cellulose molecules are solvated by solvent molecules. It appears that the electron donor–acceptor interaction must reach a certain magnitude before the destruction of the ordered regions of polymer occurs. It has been pointed out by Golova *et al.* [60] that for effective interaction with cellulose an organic solvent should possess favourable polar, basic and steric properties. Donor dipolar aprotic solvents dimethyl sulfoxide (DMSO), dimethylacetamide (DMAA), dimethylformamide (DMF), *N*-methylpyrrolidone (MP) and hexamethylphosphorotriamide (HMPTA) interact with cellulose only at the stage of swelling. If they are mixed with appropriate acceptor solvents then complexes are formed and interactions are sufficient for cellulose to go into solution.

A thermochemical investigation of cellulose esterification by nitric acid in organic media has been carried out various workers [61–63]. Heat effects due to interaction between cellulose derivatives and solvents are two to three times greater than those due to similar interactions with cellulose itself.

Tsvetkov [64] studied the solvation of cellulose nitrate in various solvents. It was shown that in electron-donating solvents, solvation was preferentially of nitro

Table 3.1 Enthalpies of mixing (ΔH_{mix}) of cellulose with solvents (mass ratio 1:40) at 298 K

Solvent	$-\Delta H_{mix}$ / kJ mol^{-1}	Solvent	$-\Delta H_{mix}$ / kJ mol^{-1}
Water	9.2	Ethylenediamine	41.8
Methanol	5.0	Hydrazine (95 mass%)	44.3
N-Methylformamide	11.3	Acetic acid	5.0
DMF	10.0	Trifluoroacetic acid	36.8
DMSO	13.4	Phosphoric acid (85 mass%)	19.2
Piperidine	1.7	Nitric acid	82.8
Monoethanolamine	15.5		

groups, and in electron-accepting solvents that of oxygen atoms in hydroxyl and acetyl groups.

The enthalpies of wetting of cellulose with water, methanol, propanol, butanol and amyl alcohol are −48.5, −30.5, −27.7, −18.9, −14.6 and −5.6 J g^{-1}, respectively, at 298 K [65]. The decrease in the enthalpies of mixing of cellulose with alcohol with increasing length of the alkyl radical of the alcohols is evidently connected with steric hindrance during solvation in solutions. The interaction enthalpies of cellulose with water, formamide, ethylene glycol, dioxane, 1,2-dichloroethane have been studied by Kuvshinnikova *et al.* [66]. It was found that the enthalpies of wetting change from −5.0 to −59.0 J g^{-1} with increasing polarity of solvent.

The electron donor–acceptor properties of all components of a system should be taken into account in an estimation of the electron donor–acceptor character of an interaction. It has been shown, on the basis of the calorimetric data, that the electron-donating ability of the polysaccharide functional groups decreases in the series chitosan > chitin > cellulose diacetate > cellulose > cellulose nitrate > monocarboxylcellulose [67]. Because of the amphoteric nature of cellulose OH groups, it can be assumed that the interaction of these macromolecules with both proton acceptors and proton donors is possible.

Golova *et al.* [60] proposed a mechanism for the solution of cellulose in solvents which do not react chemically to form a derivative in solution. They based this on data from ^{13}C NMR measurements and also assumed that interaction of cellulose with non-aqueous solvents took place by an electron donor–acceptor mechanism. It was suggested that at the first stage of the interaction between methylmorpholine-*N*-oxide (MMO) and cellulose, MMO molecules are accommodated between the crystallographic planes. In this situation, $NO—N{\rightarrow}O_{(8)}$ and groups of MMO molecules are at a distance less than 3 Å from the cellulose hydroxyls, in particular the primary ones, $C_{(6)}OH$, i.e. at a distance appropriate for the formation of hydrogen bonds. In an MMO monohydrate molecule, the semi-polar bond $N{\rightarrow}O$ has high electron donicity. This is likely to lead to the destruction of intermolecular bonds in cellulose $OH_{(6)} \cdots OH_{(3')}$ and to the formation of $OH_{(6)} \cdots O_{(8)}$ and $OH_{(3')} \cdots O_{(8)}$ bonds in the system cellulose–methylmorpholine-*N*-oxide.

The possibility of the interaction between MMO molecules and hydroxyl groups that are connected with intramolecular $OH_{(2')}—OH_{(6)}$ and $O_{(5)}—OH_{(3')}$ bonds increases with the destruction of intermolecular hydrogen bonds. The solvation of the intermolecular and intramolecular hydrogen bonds of cellulose would inevitably result in the transition of cellulose macromolecules into solution.

Investigations of the rotary and translatory mobility of nuclei of low-molecular-mass probes in non-aqueous solutions of cellulose derivatives with different degrees of substitution are discussed in Chapter 4. This work is directly connected with the investigation of the thermodynamics of the solvation of cellulose and its derivatives.

Calorimetry and IR spectroscopy were used by Zenkov *et al.* [68] to investigate the interaction between dry cellulose and liquid and gaseous phases of water. It was shown that the interaction of cellulose with water was characterized by the existence of the two stages, endothermal (quicker and less energy consuming) and then exothermal (determining the total heat effect). The authors supposed that the interaction enthalpy of cellulose with water included at least three contributions: the endothermal effect of the destruction intra- and intermolecular hydrogen bonds of cellulose on sorption of water; the exothermal effect of cellulose hydration and the exothermal effect of the hydration of free hydroxyl groups of cellulose, formed on the destruction of hydrogen bonds. The increase in the exothermal effect with increasing temperature is explained by the destruction of H-bonds of cellulose followed by hydration.

The formation of donor–acceptor complexes in solutions of cellulose and its derivatives [59, 69] has a considerable influence on the enthalpies of interaction of these polymer macromolecules with solvents. Tager [70] has observed that, as far as dissolution of polymers is concerned, the familiar generalization 'like dissolves like' should be replaced by 'the best mixing is observed between substances possessing opposite properties.' Proton-donating substances mix better with proton-accepting substances. However, one should not identify the solvent solvency or other thermodynamic characteristics of a system with any one of the physico-chemical parameters. For example, the dipole moment reflects the distribution of the electronic cloud in a given functional group but it does not enable one to estimate the dispersion interaction. The Hildebrand solubility parameter is an integral characteristic of an individual substance, but it does not take into account the specificity of functional groups or of H-bond formation between the components of a system. It is better to use three-component Hansen parameters [71] for predictions of polymer solubility. The individual components take into account the ability to form H-bonds, polar interactions and non-polar dispersion interactions. Electron donor–acceptor ability of the interacting substances can be the most important characteristic of the solution process, but insufficient experimental material and absence of quantitative data on donicity and acceptability of solvents and their mixtures make the estimation of this characteristic difficult.

The limited character of simple correlations of a solvent property, such as solvency, with its thermodynamic parameters has resulted in the development of more complicated multiparametric models [26]. The choice of multiparametric model depends on which physico-chemical properties of solvent are important in the solution process. For the systems in which the interactions have a definite donor–acceptor character it is expedient to use the Gutmann–Mayer model [26]:

$$\Delta G^{S} - \Delta G^{R} = a(\mathrm{DN}^{S} - \mathrm{DN}^{R}) + b(\mathrm{AN}^{S} + \mathrm{AN}^{R}) + c(\Delta G^{\circ S}_{\mathrm{evap}} - \Delta G^{\circ R}_{\mathrm{evap}}) \quad (3.13)$$

where ΔG^{S} and ΔG^{R} are the standard free energies of the reaction investigated in various solvents S and reference solvent R; DN, AN and $\Delta G^{\circ}_{\mathrm{evap}}$ are the values of the donor number, acceptor number and standard evaporation energy, respectively.

The Krigovsky–Fosett model [26] is applicable to describe solution processes in which the entropy changes are small, i.e.

$$\Delta S = \Delta S_0 + LE_T + \beta \mathrm{DN} \quad (3.14)$$

where E_T is the Dimroth–Reichardt electrophilicity parameter and L and β are coefficients. If a particular multiparametric model does not describe a process satisfactorily, then additional parameters relating to the peculiarities of a given system need to be added.

The existing information on the structural–thermodynamic characteristics of solutions of cellulose and its derivatives [72] shows that the enthalpic contribution is more important than the entropic contribution at low polymer concentrations. An approach based on the prediction of the solvency of solvent or solvent mixture for a definite polymer [30] can form the basis of models of solution enthalpies.

Huyskens and Hunlait [30] considered a series of factors determining polymer solubility: its phase state, chemical nature, existence of H-bonds between polymer and solvent and the differences in the dimensions of solvent and solute molecules.

When thermochemical data for the dissolution of cellulose and its derivatives is interpreted or predicted, it is important to take into account the properties of both the solvent and the polymer. Systematic information is needed on the effect of the various factors which affect the thermochemical characteristics of the process. These factors include the nature of the substituents in the cellulose molecule, the degree of replacement of hydroxyl groups and the various physico-chemical properties of the solvents.

5 INFLUENCE OF THE DEGREE OF SUBSTITUTION OF HYDROXYL GROUPS OF CELLULOSE MACROMOLECULES BY ESTERIC GROUPS AND OF THE PHYSICO-CHEMICAL PROPERTIES OF SOLVENTS ON THE SOLUTION ENTHALPIES OF CELLULOSE

The experimental determination of the enthalpies of solution of cellulose ethers is experimentally difficult owing to the long period needed for the establishment of equilibrium of the solution processes. To find the thermochemical parameters of solvation of hydroxypropylcellulose (HOPC) and other cellulose ethers it is convenient to introduce a third component, a liquid non-electrolyte as the 'probe' of the solvation processes, into a binary polymer–solvent system. The introduction of electrolytes and non-electrolytes as 'probes' generally gives good results when studying solvation processes in liquids [73–75].

Cellulose ethers contain hydrophobic pyranose rings and hydrophobic substituents as well as non-substituted OH groups. They can be considered as diphilic polymers. Meerson *et al.* [75] analysed the influence of alkyl substituents in the cellulose ethers on the properties of their aqueous solutions. Cellulose ethers

Table 3.2 Characteristics of the samples of cellulose ethers

Polymer	Groups %			
	Methoxyl	Oxypropyl	Oxyethyl	DS
Methylcellulose (MC)	2.65	26.8	—	1.6
Methyloxypropylcellulose (MOPCI)	2.60	25.1	1.5	1.5
Hydroxypropylcellulose (HOPCI)	2.40	—	32.0	0.9
Oxyethylcellulose (OEC)	1.8	—	35.1	0.8
Oxyethyloxypropylcelluose (OEOPC)	1.30	3.0	32.1	0.7
Hydroxypropylcellulose (HOPCII)	1.45	—	34.9	1.0
Methyloxypropylcellulose (MOPCII)	2.60	29.0	4.0	1.7
Methyloxypropylcellulose (MOPCIII)	0.80	27.0	1.7	1.6

were grouped into two series: methyl-, methyloxypropyl- and oxypropylcellulose (I) in one series and oxyethyl-, oxyethyloxypropyl- and hydroxypropylcellulose (II) in the other series. The series are characterized by approximately the same viscosity, but different compositions, as shown in Table 3.2.

The technique described by Meerson *et al.* [75] has been used to find the integral heat of solution of KCl in aqueous solutions of cellulose ethers of different concentrations. The results are shown in Figures 3.3 and 3.4. The figures show that all the systems studied are characterized by the existence of two minima and two maxima in the endothermicity of solution of the salt.

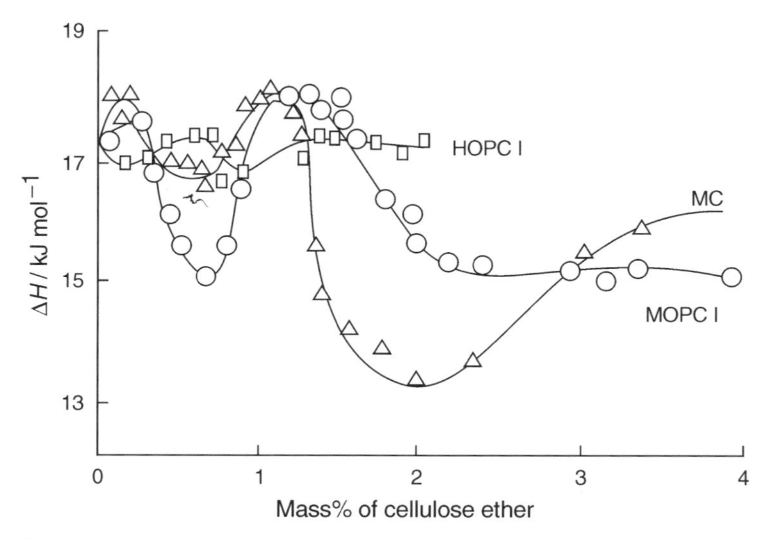

Figure 3.3 Enthalpy of solution of KCl in aqueous solutions of cellulose ethers at 299 K. See Table 3.2

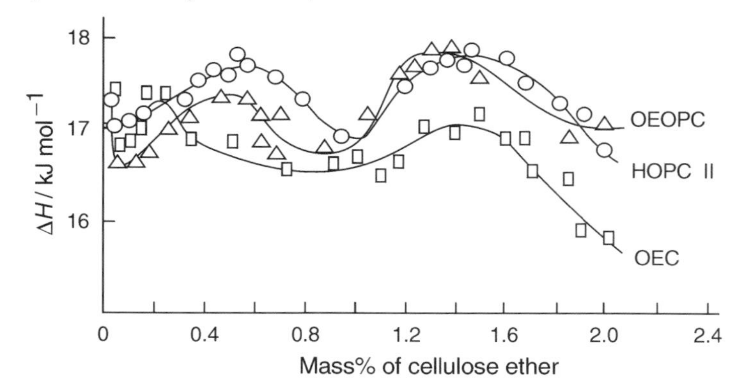

Figure 3.4 Enthalpy of solution of KCl in aqueous solutions of cellulose ethers at 299 K. See Table 3.2

The first maximum in the region 0.15% of MOPCI was interpreted as being due to stabilization of the water structure by methoxyl groups. These groups are small enough to fit into vacancies in the water structure. These groups inhibit thermal motion of water molecules by steric hindrance and by dispersion interaction. This favours hydrogen bond formation between polymer and water molecules. This phenomenon can be considered as hydrophobic hydration of non-polar groups of the polymer [76].

The increase in the concentration of MORCI results in a minimum in the curve corresponding to $\Delta H^{\circ}_{\text{soln}} = f(C)$. This is a consequence of decreasing hydrophobic hydration. In this concentration range the stabilizing effect is evidently outweighed by the effect of destruction of the water structure.

A further increase in MOPCI concentration leads to a second endothemicity maximum in the concentration region 1.2–1.5%. Presumably the stabilization of the spatial net of water hydrogen bonds is repeated.

If the data for the integral heats of KCl solution in the series MC–MOPCI–HOPCI are compared with those for the series OEC–OEOPC–HOPCII, it is seem that if methoxyl radicals are absent then the first endothermicity maximum in the low-concentration region disappears. This is replaced by an endothermicity minimum related to the destruction of the water structure due to the introduction of more hydrophilic groups. The second peculiarity of aqueous solutions of members of the HOPCII–OEOPC–OEC series is that an increase in the degree of radical hydrophilicity results in the disappearance of the water structure stabilization regions. Solutions of HOPCII are characterized by two water stabilization regions, OEOPC by a single region (1.3–1.5%) and OEC by the absence of stabilization. The presence of hydroxypropyl groups in the macromolecules of cellulose ethers favours the stabilization of water structure. This effect grows with increasing degree of substitution on hydroxypropyl groups. In addition,

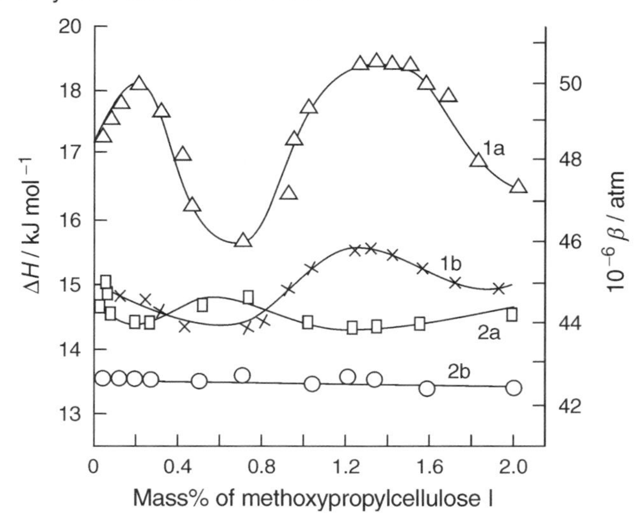

Figure 3.5 Influence of temperature on the enthalpy of solution of KCl in solutions of MOPC I in water (1a, 299 K; 1b, 319 K) and on the adiabatic compressibility of MOPC I solutions (2a, 299 K; 2b, 319 K)

the degree of substitution and the molecular mass of the samples of HOPC influence the integral heats of solution of KCl [79].

Figure 3.5 shows the influence of temperature on the integral enthalpy of solution of KCl, ΔH°_{soln}, when it is dissolved in water containing various concentrations, C, of MOPCI. The figure also shows variation of adiabatic compressibility with variation of concentration. The curves showing change of ΔH°_{soln} with concentration at 294 and 314 K indicate that with increasing temperature the curve shifts less positive values of ΔH°_{soln} for the salt. This is related to the destruction of the water structure.

Myasoedova *et al.* [77] have studied the dependence of heats of solution of cellulose and its acetates on the degree of substitution of hydroxyl groups by acetate groups in non-aqueous single and mixed solvents at 298 K. For all the solvents studied, there is a linear correlation between the values of the enthalpies of solution and the amount of bonded acetic acid in the cellulose acetates. Hypothetical values of enthalpies of solution of cellulose were obtained by extrapolating the linear dependence of ΔH°_{soln} on the degree of substitution to zero substitution for solutions of cellulose acetates in trifluoroacetic acid and in its mixtures with chlorinated hydrocarbons [77].

Extrapolated and experimental values of ΔH°_{soln} for cellulose in CF_3COOH and its mixtures with chlorinated hydrocarbons are close in value. The difference is not higher than 15 J g^{-1}. This difference is due to the differing in degrees of crystallinity of samples of wood cellulose and cellulose acetates. The additional endothermal contribution to ΔH°_{soln}, due to melting of the crystalline regions of cellulose, was estimated to be 15 J g^{-1} from literature data on the heats of melting of cellulose [54] and the degrees of crystallinity. This confirms that the linear variation of ΔH°_{soln} with degree of substitution extends to zero substitution. This relationship provides a method of determining the degree of substitution of cellulose acetates by calorimetry.

The experimental data which are available enable one to estimate the influence of the nature and composition of the solvent on ΔH°_{soln} of cellulose acetates with various degrees of substitution. Adding chlorinated hydrocarbons $C_xH_yCl_z$ to trifluoroacetic acid results in an increase of solvency for cellulose and its acetates [77]. This is accompanied by an increase in absolute values of ΔH°_{soln} in the order $CF_3COOH < CF_3COOH + CHCl_3 < CF_3COOH + CH_2Cl_2 < CF_3COOH + C_2H_4Cl_2$. This can be explained by the influence of chlorinated hydrocarbons in reducing self-association of the acid [78]. The increase in reactive monomers of the acid leads to an increase the acceptor properties of the mixtures with respect to cellulose polymers.

The increase in exothermicity of the cellulose acetates is observed over the series of individual solvents: DMAA < DMF < DMSO < CF_3COOH. The ability of a solvent to solvate cellulose depends on both its donor and its acceptor numbers.

The importance of specific solvation in the dissolution of cellulose acetates is clearly illustrated by the linear dependence of ΔH°_{soln} on the half-sum of the donor and acceptor numbers, $\frac{1}{2}$(DN + AN), for cellulose diacetate (CDA) and cellulose triacetate (CTA) in the solvents studied [79]. This linear correlation applies to measurements of ΔH°_{soln} for cellulose acetates in the solvents of widely different nature with considerable differences in donor and acceptor properties, from proton-donating CF_3COOH to dipolar aprotic DMAA, DMF and DMSO. High values of the heat effects of solution of cellulose and its derivatives in non-aqueous solvents are due to the formation of hydrogen bonds between polymers and solvents. This is confirmed in the case of cellulose acetate in DMSO by ^{13}C nuclear magnetic relaxation measurements.

Investigation of solutions of cellulose acetate with different degrees of substitution in deuterated DMSO by molecular NMR relaxation probes is discussed in Chapter 4. There is a linear correlation between the parameter of spin–lattice relaxation, *W*, and the degree of substitution of the cellulose acetates. There is also a linear correlation between ΔH°_{soln} and the value of *W* for different cellulose acetates (Figure 3.6). This dependence reflects the influence of the change in the degree of substitution on energetic and structural parameters during the formation of polymer–solvent hydrogen bonds.

The degree of substitution of hydroxyl groups in cellulose by acetate groups influences the coefficients of translatory diffusion of molecular probes in a similar way [80]. In this case the characteristics of the translatory motion of the probes are more sensitive to the supramolecular structure of the polymer solutions than are the characteristics of their rotary mobility.

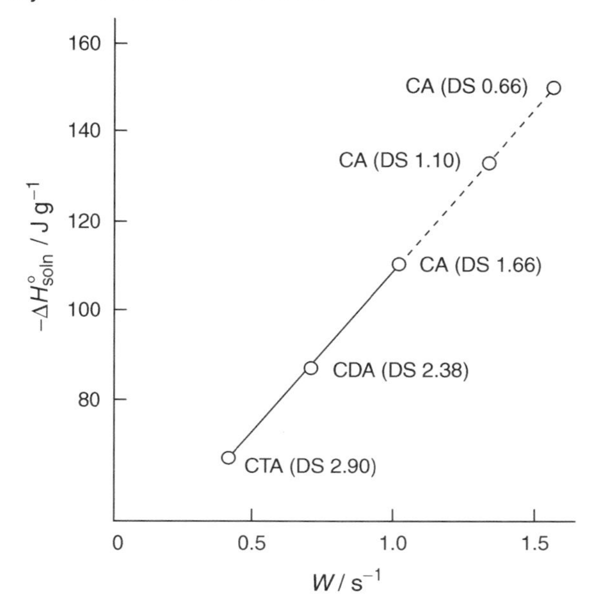

Figure 3.6 Relationship between the enthalpy of solution in DMSO of cellulose acetates with different degree of substitution and the spin–lattice relaxation of DABO in the solution

The maxima in the variations with concentration of cellulose acetates of the correlation times of rotary motion of groups in aqueous solution, calculated from NMR spectra, occur in the concentration region in which the enthalpy of dissolution of the salt reaches a second maximum. In addition, the highest solubility of benzene in the aqueous solutions of ethers discussed above, as well as the maxima in the refractive indices also occur in this region. These confirm the hypothesis [80] that three-dimensional fluctuation nets of hydrogen bonds are formed by hydrophobic interactions in this region.

The author and her colleagues have experimentally measured the enthalpies of solution of DMSO in HOPC solutions, water, ethanol, dimethylacetamide, dimethylformamide and dioxane (DO). The results of the thermochemical investigations are shown in Figure 3.7. As can be seen, the concentration of HOPC does not appreciably influence the values of ΔH°_{soln} of DMSO in hydroxypropylcellulose solutions. In addition, Figure 3.7 shows the increase in the values of ΔH°_{soln} for DMSO over the series of solvents in the order water < DMAA < DMF < DO < ethanol.

Enthalpies of solution of DMSO in ethanol, DMAA and DMF have also been determined. Values of ΔH°_{soln} for DMSO in water and DO agree well with those from the literature [53, 80]. Mixing of two donor liquids is known to be accompanied by an endo-effect [81]. It is noteworthy that such an effect also occurs

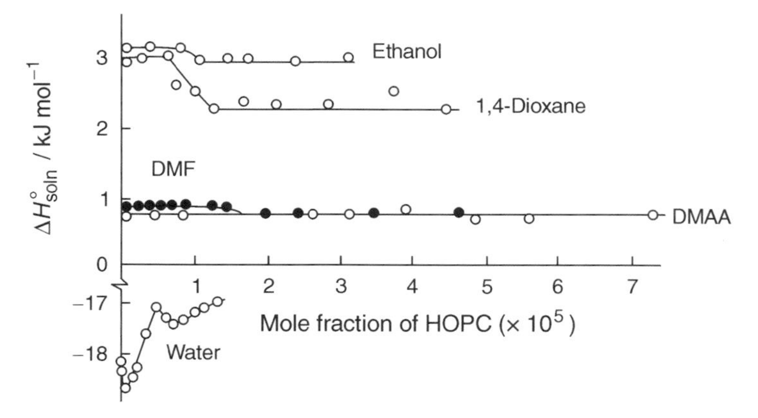

Figure 3.7 Variation of the integral enthalpy of solution of DMSO with the mole fraction of HOPC in various solvents

on mixing DMSO with ethanol. This phenomenon can be explained by the change in ethanol donicity in the associated state, as indicated by values of donor numbers, DN: $DN_{SbCl_5} = 18.5$, $DN_{bulk} = 27.8$ according to Marcus [82] and $DN_{bulk} = 32$ according to Gutmann [19].

The determination of ΔH°_{soln} of DMSO (solvent I) in individual solvents II was the first stage in the investigation on the regularities of solvation in solvent I–solvent II–polymer systems. On going from DMSO–solvent II solutions to HOPC–solvent II systems, the value of ΔH°_{soln} of DMSO depends on the processes of polymer solvation in solution. It is known that at a concentration of cellulose and its derivatives above 1 mass%, there occurs an interaction between the polymer molecules to form a fluctuation net of hydrogen bonds [72, 75]. In the systems noted, this occurs in the concentration region of HOPC solutions, $x_{HOPC} > 125 \times 10^{-7}$, where the enthalpy of solution of DMSO does not change with increasing polymer concentration.

The extreme variation of ΔH°_{soln} with concentration of HOPC in water is evidently connected with the change in the structure of the solutions. The increase in the exothermicity of solution of DMSO in an aqueous solution of HOPC at mole fraction 7.5×10^{-7} is a result of the destruction of the water structure by polymer molecules. A further increase in HOPC concentration results in the stabilization of the aqueous solution.

In the concentration region of the second maximum of ΔH°_{soln}, there are several significant factors. There is simultaneous strengthening of intermolecular hydrogen bonds between molecules of the dissolved polymer. This results in solvation of the dissolved HOPC molecules by additional water molecules. The general tendency towards a decrease in exothemicity of the interaction with DMSO with increasing HOPC concentration in aqueous solution is probably due to the stabilization of the solution structure by the formation of cooperative hydrogen bonds of molecules HOPC–HOPC, HOPC–water.

Figure 3.7 shows that in non-aqueous solutions up to a mole fraction of about 1.25×10^{-5} of HOPCI, the value of ΔH°_{soln} for DMSO remains constant, within the limits of experimental error. Taking into account that ethanol, DO, DMF and DMAA are associated liquids, one can assume that the primary structure of solvent is preserved after small additions of polymer.

It is convenient to use the transfer enthalpy of a substance from one solvent into another as the thermochemical characteristic of solvation of atomic or molecular particles in a series of individual solvents [17]. The transfer enthalpies (ΔH°_{tr}) of DMSO from an individual solvent into concentrated polymer solution are appropriate thermochemical characteristics that are sensitive to the solvation processes.

Statistical methods were used to reveal the interrelation between the physico-chemical properties of solvents and experimental thermochemical parameters, ΔH°_{soln} and ΔH°_{tr}, for transfer of DMSO from solvent to HOPC solutions [83]. Parameters of some solvents analysed in this way are given in Table 3.3.

Pairwise and multiple regression analysis of ΔH°_{tr}, $\Delta H^\circ_{tr}/\Delta H^\circ_{soln}$ and the physico-chemical properties of solvents has been carried out. The advantages of using absolute rather than relative values of the heats of transfer to estimate the solvent influence are evident when comparing the correlations of ΔH°_{tr} and

Table 3.3 Some physico-chemical parameters of solvents for HOPC

	Solvent					
Parameter [a]	Water	Ethanol	DMAA	DMF	1,4-Dioxane	Ref.
$\frac{n^2-1}{n^2+2}$	0.2054	0.2214	0.2627	0.2584	0.2543	118
$\frac{\varepsilon-1}{2\varepsilon+1}$	0.490	0.470	0.480	0.480	0.223	118
B	—	235	343	291	237	120
E_T	264.1	217.3	183	183.4	150.7	120
z	396	333.2	280	286.7	270.4	120
DN_{SbCl_5}	18	19.2	27.8	26.6	14.8	19, 26
AN	54.8	37.1	13.6	16.0	10.8	19, 26
π^*	1.09	0.54	0.88	0.88	0.55	27
ε	78.3	25.2	37.8	36.7	2.21	119
μ/D	1.84	1.69	3.81	3.86	0.4	119
$\overline{V}/cm^3\,mol^{-1}$	18.07	58.69	92.98	81.12	85.71	119
$\rho/g\,cm^{-3}$	0.997	0.785	0.937	0.901	1.028	119
δ	98.08	54.33	46.55	50.86	41.9	71

[a] $(n^2-1)/(n^2+2)$ is polarizability; $(\varepsilon-1)/(2\varepsilon+1)$ is polarity; B is basicity; E_T is Dimroth–Reichardt electrophilicity parameter; z is Kosover electrophilicity parameter; DN, AN are donor and acceptor numbers according to Gutmann; π^* is Kamlet–Taft bipolarity parameter; ε is dielectric permeability; μ is dipole moment; $\overline{V}$ is molar volume; ρ is density; δ is three-dimensional solubility parameter.

Table 3.4 Heats of transfer (kJ mol^{-1}) of DMSO from solvent into HOPC solution: experimental and calculated values using a linear model, $\Delta H^{\circ}_{tr} = A + B\varepsilon$[a]

Solvent	$\Delta H^{\circ}_{tr\,exp}$	$\Delta H^{\circ}_{tr\,calc}$	$\Delta H^{\circ}_{tr\,exp} - \Delta H^{\circ}_{tr\,calc}$
Water	0.870	0.825	0.045
Ethanol	−0.164	−0.196	0.032
DMAA	0.021	0.046	−0.025
DMF	−0.069	0.025	−0.094
1,4-Dioxane	−0.596	−0.638	0.042

[a] $A = -0.6809 \pm 0.06691$ kJ mol^{-1}; $B = 0.01924 \pm 0.00153$ kJ mol^{-1}.

$\Delta H^{\circ}_{tr}/\Delta H^{\circ}_{soln}$. The most significant parameters that influence ΔH°_{tr} are dielectric permeability (ε), Hildebrand solubility parameter (δ), Dimroth–Reichardt electrophilicity parameter (E_T) and the sum of donor and acceptor numbers (DN + AN) of solvents. The regression coefficients decrease from 0.9 to 0.7 over the series noted (Table 3.4). Since there is a strong correlation between ΔH°_{tr} and ε, it is appropriate to use a linear regression model, i.e.

$$\Delta H^{\circ}_{tr} = A + B\varepsilon \tag{3.15}$$

Coefficients of this model are given in Table 3.4. The comparison between the calculated results and experimental data (Figure 3.8) shows that the errors of approximation (3.16) do not exceed 0.094. The criterion of the linear approximation, F, is the ratio between dispersion to the average ΔH°_{tr} value and dispersion with respect to the calculated values; the value of F characterizes the quality of the model proposed and has a value of 39.9. Since the decrease in the pairwise correlation coefficients is observed for the other physico-chemical parameters, they cannot be used to build a single-parametric linear model similar to Eq. (3.15). A model has been constructed to enable the precision of calculation to be increased. Use was made of methods of multiple regression and singular analysis of the redetermined system of equations [84]. These enabled the best approximate relationship to be found [53, 85]:

$$\Delta H^{\circ}_{tr} = A_0 + A_1X_1 + A_2X_2 + A_3X_3 \tag{3.16}$$

where

$$X_1 = \frac{\varepsilon - 40}{30}; \quad X_2 = \frac{E_T - 50}{14}; \quad X_3 = \frac{(\mathrm{DN} + \mathrm{AN})/2 - 24}{12}$$

Coefficients of this equation and comparison between the calculated results and experimental data are shown in Table 3.5. The criterion of the model, F, equals 292. The calculation error does not exceed 0.024. These results indicate the high precision of this multiparametric approximation. All the coefficients of Eq. (3.16) are meaningful at a probability level exceeding 95%.

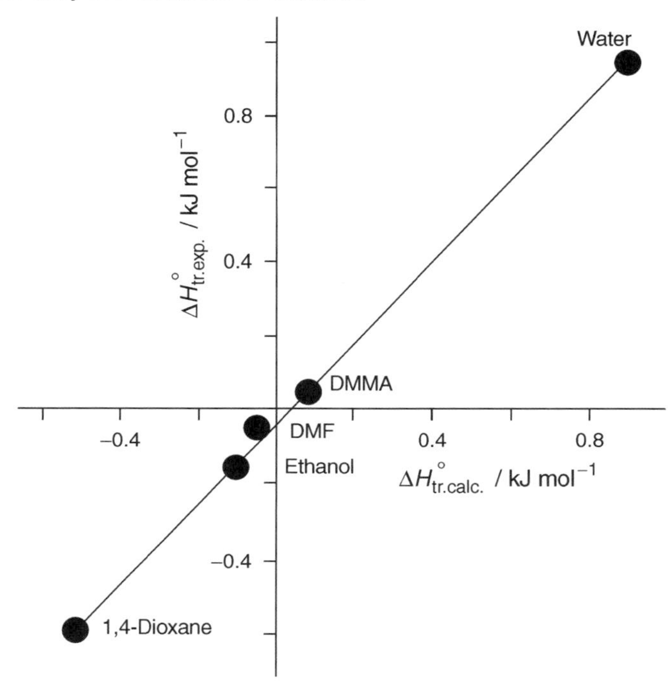

Figure 3.8 Comparison between experimental and calculated values of DMSO transfer enthalpies from solvent into HOPC solution

Table 3.5 Comparison between experimental and calculated values of the heats of transfer (kJ mol^{-1}) of DMSO from solvent to HOPC solutions[a]

Solvent	$\Delta H^{\circ}_{tr\,exp}$	$\Delta H^{\circ}_{tr\,calc}$	$\Delta H^{\circ}_{tr\,exp} - \Delta H^{\circ}_{tr\,calc}$
Water	0.870	0.874	−0.004
Ethanol	−0.164	−0.168	0.004
DMAA	0.021	0.024	−0.003
DMF	−0.069	−0.025	−0.044
1,4-Dioxane	−0.596	−0.591	-0.005

[a] $A_0 = 0.1953 \pm 0.0396$ kJ mol^{-1}; $A_1 = 0.6517 \pm 0.0446$ kJ mol^{-1}; $A_2 = 0.7687 \pm 0.2491$ kJ mol^{-1}; $A_3 = -0.6753 \pm 0.2517$ kJ mol^{-1}.

The model developed is similar to the Krigovsky–Fosett model [26]. The latter can be used to describe the influence of the solvent on solvation thermodynamics characterized by large contributions of enthalpic factors to the energetics of the processes.

Analysis of the coefficients of Eq. (3.16) shows that the largest contribution to the enthalpy characteristic of DMSO solvation in the HOPC–solvent I–solvent II system is due to the dielectric properties of non-aqueous solvents. This agrees well with Solomonov's hypothesis [23] about the interrelation between the enthalpy of non-specific solvation of organic non-electrolytes and mole refraction of solvents. It also shows that both the non-specific (dipole–dipole, dispersion) and the specific (hydrogen-bonding) constituents contribute to the entropy of solvation of cellulose derivatives by non-aqueous solvents.

A special feature of the proposed model is the need to take into account both the donor and acceptor numbers of non-aqueous solvents when the specific solvation of hydroxyl groups of cellulose macromolecules is considered. These groups possess both donor and acceptor properties.

6 INFLUENCE OF THE DEGREE OF SUBSTITUTION OF HYDROXY GROUPS OF CELLULOSE MACROMOLECULES BY ACETATE GROUPS AND OF POLYMER CONCENTRATION ON THE ENTHALPIES OF SOLUTION OF POLYETHYLENE GLYCOL IN SOLUTIONS OF CELLULOSE ACETATES IN DIMETHYL SULFOXIDE

Zavyalov *et al.* [86] have measured the enthalpies of solution, ΔH°_{soln}, of polyethylene glycol (PEG) with $M = 400$ in solutions of cellulose acetates in DMSO at 298 K. The degrees of substitution (DS) of the cellulose acetates were 0.63, 0.96, 1.55, 2.43 and 2.90 and the polymer concentration range 0–4 mass% of PEG. Dissolution of PEG in the systems studied is characterized by negative ΔH°_{soln} values. The different ways in which enthalpies of solution change with concentration is related to changes in the character of the solutions. For solutions of cellulose acetates with DS $= 0.63$, there is no appreciable change in ΔH°_{soln} for PEG with change in CA concentration. With increasing DS of CA, there is an increase in the exothermicity of solution of PEG in the concentration range 1–2 mass%.

The dependence on the degree of substitution, of the initial concentration, C_{in}, at which the increase in absolute values of ΔH°_{soln} of PEG in CA solutions occurs has also been studied [86]. There is a tendency for C_{in} to decrease with increasing degree of substitution of CA. This is probably due to the formation of a fluctuation net of hydrogen bonds, formed as a result of the solvation process and the breaking of intra- and intermolecular bonds of CA.

The dependence the enthalpy transfer of PEG from DMSO into CA solution on the degree of substitution, DS, at concentration C_{in} has also been studied [86]. In the $0.6 < \mathrm{DS} < 1.0$ region, there is a very sharp increase in the exothermicity of transfer. This is associated with a strengthening of PEG–CA intermolecular interactions. At DS ≥ 1.0, absolute ΔH°_{tr} values of PEG are practically independent of the degree of substitution. Under these conditions the replacement of hydroxyl groups by acetate groups causes little change in the interaction between PEG and the cellulose derivative. For CA with DS ≤ 0.63, the PEG–CA interaction is complicated. It is likely that the strengthening of the solvation of CA macromolecules with increasing number of non-substituted OH groups of polymer is one possible reason for this.

7 CHEMICAL HETEROGENEITY AND THERMOCHEMICAL PROPERTIES OF CELLULOSE

Modern concepts of the cellulose structure have been described in various monographs [87–90]. The influence of peculiarities of the cellulose structure and of chemical heterogeneity on the thermochemical properties has been discussed. It has been shown that links containing carboxyl and aldehyde groups are present in natural samples of wood and cotton cellulose together with the glycopyranose cycles. They are formed in oxidation processes by the action of oxygen in the air. The acidic character of unpurified cellulose preparations is due principally to the presence of residues of polyglucuronic acid, which is a constituent of pectin substances.

Oxidation of cellulose occurs not only during biosynthesis, but also during extraction and storage. Residues of pitches and fats, together with isomeric structures such as mannan, galactan and oxidation products, can be found in poorly purified samples of cellulose. In addition lignin, a phenylpropane polymer containing oxygen, may be present in wood cellulose [91].

It is interesting to consider the change in the heat of combustion of cotton cellulose with change in the degree of purification by boiling and bleaching [87]. It is only after the second stage of purification that the heat of combustion of cotton fibre approaches that of pure cellulose, 17.56 kJ g^{-1}, to within 0.2%.

When experimental and calculated values of the heats of combustion of cotton cellulose are compared it is evident that they assume intermediate values between ΔH_{comb} of the initial raw material and ΔH_{comb} of cotton which has been subject to boiling with alkali. The difference between the experimental values of ΔH_{comb} and the calculated values may be due to the presence of incombustible admixtures. The presence of 0.2–1.0% of calcium, iron and other metal salts in samples of cotton cellulose samples can lower the heat of combustion by 33.47–175.43 J g^{-1} [87]. Stages in the sulfatic purification of coniferous wood cellulose also cause changes in heats of combustion which gradually approach the calculated value for pure cellulose [92, 93]. Combustion calorimetry, in combination with the model calculations, can be used to judge the degree of chemical and physical heterogeneity of a sample of cellulose.

8 RELATION BETWEEN THE STRUCTURE OF CELLULOSE AND ITS DERIVATIVES AND THE THERMOCHEMICAL CHARACTERISTICS OF DISSOLUTION

Thermochemical characteristics of solutions of cellulose and its derivatives are sensitive to structural peculiarities. Changes in the degree of crystallinity of

Table 3.6 Integral heats of interaction of natural and regenerated cellulose with water [95]

Water content in the sample/ mass %	Heat of inter-action/J g^{-1}		Water content in the sample/ mass %	Heat of inter-action/J g^{-1}	
	Cotton	Viscose fibre		Cotton	Viscose fibre
0	42.7	93.2	7	8.4	—
0.5	37.7	99.2	8	7.6	40.2
1	33.5	84.1	10	—	33.9
3	20.6	67.0	12	—	30.6
5	12.6	54.5	—	—	—

cellulose and other polysaccharides over wide ranges influence the heat effects of wetting, swelling and also dissolution of polymers in solutions containing other solutes. In this last case, additional energetic effects occur due to the destruction of the crystalline regions of cellulose and to its rubber transition under the influence of low-molecular-mass compounds.

Experimental data on the heats of wetting, swelling and interaction of cellulose with various liquids that are available in the literature vary from 40.7 to 63.0 J g^{-1} [93, 94]. The wide range of the experimental data may be due to the presence of admixtures and differences in supramolecular structure and in the humidity of samples investigated [95] (Table 3.6).

Matyushin *et al.* [96] measured the combustion enthalpies of two samples of cotton cellulose, vacuum dried for the same time but at different temperatures. An increase in the drying temperature caused an appreciable increase in the heat of combustion. This was due to the polymer losing physically bonded water. Rabek [97] found that above 150 °C, there is a decrease in the ability of cellulose to absorb and retain water due to a process of intermolecular dehydration.

Because of the strong influence of humidity on thermochemical parameters, Zharkovskii and Emelyanova [98] dried samples at 110 °C before making measurements. This completely eliminated moisture. However, drying of cellulose at temperatures > 50 °C results in thermal destruction [99] and has a noticeable influence on the porous structure [100] by causing a decrease in the internal surface of the cellulose fibres. The author and her colleagues consider that samples of cellulose should be dried at ~ 50 °C.

The relation between the amount of water bonded with a range of hydrophilic substances and the heat effect due to interaction of this substance with water is given in Ref. 101. It was assumed that there was bond formation between the glucopyranose link of the cellulose macromolecule and 1–2 water molecules.

The influence of the degree of crystallinity on the interaction enthalpy ($\Delta H_{\mathrm{interac}}$) of various cellulose samples with water has been studied by Tsvetkov *et al.* [2]. $\Delta H_{\mathrm{interac}}$ decreases linearly with increasing degree of crystallinity.

Table 3.7 Density and enthalpy of solution of various modifications of dextran

Phase state of dextran	Type of crystalline modification	Density of the preparation/g cm^{-3}		$-\Delta H_{soln}$/kJ mol^{-1}		Degree of crystallinity/%
		Non-ground	ground	In 10% NaOH	In water	
Amorphous	—	1.304	1.303	44.9	29.5	—
Crystalline	1	1.372	—	37.5	27.2	15.0
	1	1.536	1.341	34.5	9.76	28.6
	2	1.473	1.360	39.6	—	23.2

Extrapolation of experimental data indicate that for completely crystalline cellulose the interaction enthalpy is close to zero, and for completely amorphous cellulose it is 168 J g^{-1}. A consideration of the thermodynamic characteristics of the state of ordering of the cellulose supramolecular structure shows that the entropy of state is minimum in the highly crystalline parts and is maximum in the highly amorphous parts [102].

Differences in phase states affect densities of solutions and enthalpies of solutions of dextrans in water and NaOH solutions (Table 3.7). Experimental data from Ref. 103 show that dextran polymorphism is related to conditions of crystallization [104].

As in the case of cellulose derivatives, there is a decrease in the absolute values of enthalpies of solutions of dextran with increasing degree of crystallinity.

Mikhailov and Fainberg [105] showed that conformational changes influence the enthalpic characteristics of a polymer–solvent system. This is confirmed by the considerable difference in the integral heats of solutions of different samples of viscose in a 36% aqueous solution of quaternary ammonium base. The value for non-stretched viscose fibre is 168.4 J g^{-1}, that for very firm cord viscose fibre is 154.1 J g^{-1} and that for highly stable staple viscose fibre is 124.7 J g^{-1} [106, 107].

Experimental data for integral interaction enthalpies of wood cellulose and other polysaccharides with water and organic substances [108] have shown that the value of $\Delta H_{interac}$ was lower than for polysaccharides which had a disordered supramolecular structure but a similar chemical structure. It was found that $\Delta H_{interac}$ of water with cellulose was close to the corresponding value for amorphous dextran.

Interaction of amides (monoethanolamine, DMF, trimethylformamide) with cellulose proceeds with a considerably lower value of $\Delta H_{interac}$. This indicates that firm intermolecular bonds are formed between the solution components. The concentration dependence of the interaction enthalpies of mono- and polysaccharides with aqueous solutions of ethylenediamine of different concentrations is close to linear. A non-monotonous concentration dependence is observed for $\Delta H_{interac}$ of water with solutions of ethylenediamine in water. This phenomenon is explained by the authors as being due to processes of decrystallization and recrystallization of cellulose.

The influence of various types of treatment of cellulose on its interaction enthalpies with water has been reported by Zenkov *et al.* [68]. The value of $\Delta H_{\text{interac}}$ for a normal sample of cellulose is 34.8 J g^{-1}, and for an activated sample is 80.0 J g^{-1}. The increase in absolute values of $\Delta H_{\text{interac}}$ for cellulose after various kinds of treatment is due to the destruction of intermolecular hydrogen bonds. This results in the strengthening of the interaction of cellulose with water.

A method of determining the degree of crystallinity from thermochemical data has been proposed by Dole and Tsao [109]. The following correlation was suggested:

$$Q = -X\Delta H_{\text{cr}} + \Delta H_{\text{am}} \tag{3.17}$$

where Q is the heat of solution, X is the fraction of the crystalline phase in a sample, ΔH_{cr} is the enthalpy of melting of the cellulose crystalline regions and ΔH_{am} is the heat of solution of completely amorphous cellulose. The value of ΔH_{cr} was calculated using data on the heats of solution of cellulose samples with different degrees of crystallinity in various solvents. Comparison between values of enthalpies of crystallization of cellulose I ($\Delta H_{\text{cr}} = -96\,\text{J}\,\text{g}^{-1}$) and of cellulose II ($\Delta H_{\text{cr}} = -118\,\text{J}\,\text{g}^{-1}$) [109] confirms that crystallites of cellulose II have higher energetic stability than those of cellulose I [110].

Some authors [3] have suggested that the estimation of the enthalpy of melting of the crystalline regions of cellulose should be based on experimental measurements of the enthalpies of solution of levoglucosan (model compound for cellulose) over a wide concentration range. On this basis the estimated value is 87.7 J g^{-1}.

Other authors [1, 50, 59] considered the enthalpy of solution of cellulose, and other polysaccharides possessing the amorphous–crystalline structure, to be an integral value made up of several contributions associated with different stages in the solution process.

Basedow *et al.* [50] postulated that the solution process of polymers in liquids can be assumed to be in three stages. The first is the transition of a solid polymer to a hypothetical liquid state. The second is the solution and solvation of the polymer molecules. The third is the process of mixing of the dissolved polymer molecules with solvent up to infinite dilution. The integral enthalpy of solution, ΔH_{soln}, can be expressed as follows:

$$\Delta H_{\text{soln}} = \Delta H_{\text{cr}} + \Delta H_{\text{gel}} + \Delta H_{\text{interac}} + \Delta H_{\text{mix}}. \tag{3.18}$$

The first stage is characterized by the enthalpy of crystalline lattice destruction (ΔH_{cr}) and enthalpy of transition into a gel-like state (ΔH_{gel}). The second stage includes bond formation and strong interaction ($\Delta H_{\text{interact}}$) between molecules of polymer and solvent. The third stage combines all thermal effects (ΔH_{mix}) occurring on the conformational changes of the polymer molecules during dilution.

Meerson and Lipatov [1] proposed a similar correlation:

$$\Delta H_{soln} = \Delta H_{mix} + \Delta H_{cr} + \Delta H_{gel} \quad (3.19)$$

Tsvetkov [59] gave the following expressions to estimate the interaction enthalpies of cellulose with solvents:

$$\Delta H_{interac} = \Delta H_{cr} N_{cr} + \Delta H_{gel} N_{am} \quad (3.20)$$

for the systems in which the crystalline regions are not destroyed and

$$\Delta H_{interac} = \Delta H_{cr} N_{cr} + \Delta H_{gel} N_{am} + \Delta H_{mix} \quad (3.21)$$

for the systems in which the crystalline regions are destroyed. In these expressions N_{am} is the mole fraction of the amorphous glass-like part of cellulose and N_{cr} is the mole fraction of the crystalline regions of cellulose.

Different authors have proposed similar approaches to the estimation of the interaction enthalpies of cellulose and other polysaccharides with solvents. All the above expressions take into account the amorphous–crystalline state of cellulose and the conformational changes during dilution of the polymer solution. However, it is noteworthy that determination of the enthalpies of interaction of cellulose and its derivatives with solvents [2, 68] was carried out in heterophase systems. That is why the attribution of the heat effects of the processes studied to a definite thermodynamic equilibrium state was uncertain. The data for enthalpies of solution of cellulose and its derivatives allow a more precise determination of the energetics of intermolecular interactions. Assumptions which are necessary when estimating enthalpies of melting of crystalline regions and enthalpies of mixing of cellulose with solvent are not necessary in this case [3].

It is necessary to define a general standard state when comparing thermodynamic data for solutions of cellulose and its derivatives. In principle, the choice of the standard state is determined by the stability of substances under standard conditions and by the thermodynamic parameters of the formation of chemical compounds from monatomic gases [8]. Since polysaccharides are non-volatile, the choice of an ideal gas as standard state is excluded. Pure crystalline mono- and polysaccharides are not appropriate for this purpose. In this case, numerous crystallographic factors determine the behaviour of these substances as they dissolve. It has been suggested [111] that the system 6C + 12H + 6O should be accepted as the standard state for saccharides. However, data on the heats of combustion of pure crystalline sugars are necessary in this case. The absence of reliable thermodynamic data for combustion of sugars means that the problem cannot be unambiguously resolved.

9 DEPENDENCE OF THE ENTHALPIES OF SOLUTION, DILUTION AND INTERACTION ON THE MOLECULAR MASS OF POLYMERS

One of the problems concerning the thermodynamics of solution of polymers is how molecular mass influences heat effects. Dextrans are polysaccharides. Insights can be gained by comparing their thermochemical properties with those of cellulose and relating this comparison with their structures.

The enthalpies of solution, dilution and mixing of dextrans with different molecular masses were determined by Basedow *et al.* [50] and by Myasoedova *et al.* [51]. The dissolution of dextran in all the solvents which have been studied is highly exothermic owing to hydrogen bond formation.

Myasoedova *et al.* [51] found that enthalpies of solution, ΔH_{soln}, of saccharides in water and DMSO increase with increasing molecular mass (M) up to a degree of polymerization of 6 ($M < 1000$). The similar form of the relation $\Delta H_{\mathrm{soln}} = f(M)$ for both water and DMSO has been explained as being due to changes in the morphology of saccharides with increasing M [50].

In general, solvents considerably influence the way in which ΔH_{soln} varies with increase in M. For the saccharides with $M > 1000$, a small increase in ΔH_{soln} with increase in M is observed. Taking into account that dextrans with high molecular masses are completely amorphous, the increase in ΔH_{soln} can be explained by an increase in the enthalpy of the transition into a gel-like state with increasing M.

Information on dilution enthalpies of polymers throws light on polymer–solvent interactions. The enthalpies of dilution in water and DMSO of dextrans with molecular mass < 2000 decrease markedly with increasing molecular mass. It has been suggested that dextrans of low molecular mass exist as rod-like oligomers but when $M > 2000$ they exist as spirals. When $M > 2000$ the experimental values of enthalpies of dilution are very low and independent of the molecular mass. In this case very small changes may be taking place in the spiral conformation of dextran during dilution. It is also possible that association–dissociation phenomena occur when the dextran molecules are dissolved. It is difficult to draw conclusions about thermodynamic properties of polymer–solvent systems directly from the enthalpies of dilution. It is, however, possible to calculate enthalpies of mixing from enthalpies of dilution.

Systematic investigations by Basedow *et al.* [50] show that hydrated dextran has a relatively compact spiral-like conformation in solution with no association in dilute and moderately concentrated solutions. The conformational state of dextrans in water and DMSO is similar even though DMSO is a better solvent than water.

Some cellulose derivatives show regular changes in thermochemical properties with change in molecular mass, e.g. cellulose nitrates, cellulose triacetate and octaacetylcellobiose [112]. The heats of solution in acetone of nitrocellulose with molecular masses 16 600, 23 000 and 40 000 are all equal to $-80.8\ \mathrm{J\ g^{-1}}$. The heat of solution of cellulose triacetate in acetone and chloroform is higher than that for octaacetylcellobiose in each case. Investigations of the influence of molecular mass on the thermochemical characteristics of interaction of four samples of cellulose triacetate with chloroform and acetone [113] show an increase in the heats of polymer interaction with solvent with increasing molecular mass of the polymer. However, the authors did not take into account the difference in the degrees of substitution of the cellulose triacetate samples (from 0.60 to 0.65) which would inevitably influence the solution enthalpies [114]. In addition, the authors [113, 114] did not carry out a comparative analysis of the crystallinity of the samples, i.e. the comparison was made without taking into account the differences in structural–chemical parameters of the samples investigated. Ambiguity in the interpretation of experimental results is a common fault of work in this period on the thermochemistry of cellulose and its derivatives.

A linear correlation between values of the molecular masses of different cellulose samples and values of their interaction enthalpies with an 8.7% solution of NaOH has been published by Sokolov *et al.* [115].

The interaction enthalpies for interaction with sodium tartrate of ethanol, ethylene glycol, glycerine, diethylene glycol, α-methylglucoside and cellobiose are 5.56, 9.20, 11.72, 19.25, 23.22 and 54.02 kJ mol^{-1}, respectively [116]. These hydroxyl-containing compounds are often taken to be models for the behaviour of cellulose. There is a general tendency for the enthalpy of interaction to increase with increasing number of hydroxyl groups in the molecule.

According to Meerson and Lipatov [1], there is an increase in the heat of solution with increasing molecular mass for polymers with rigid chains which cannot be packed densely.

Basedow *et al.* [50] interpreted the increase in the enthalpies of solution of oligomers ($M < 1000$) in water and DMSO in terms of the influence of the molecular mass. However, the degree of crystallinity decreases with transition from glucose to oligosaccharides. This could be the main cause of the changes in the enthalpies of solution.

Small amounts of crystalline phase [2], traces of moisture [95] and differences in the degree of substitution [114, 117] have a direct influence on the enthalpy of solution. It follows that complete information on the morphology of the substances investigated is necessary before one can use experimental data to deduce the influence of the molecular mass of cellulose and its derivatives on $\Delta H_{\text{interac}}$.

Completely unambiguous measurements of enthalpy changes during solution processes of cellulose and its derivatives are not available. As a consequence, the prediction of enthalpies of solvation from the multiparametric model discussed earlier in the chapter is especially important.

REFERENCES

1. Meerson S. I., Lipatov S. M. *Zh. Vses. Khim. Ova.* **1961**, *6*, 412–16.
2. Tsvetkov V. G., Ioyelovich M. Ya., Kaimin I. F., Reisinsh R. E. *Khim. Dreves.* **1980**, *5*, 12–15.

3. Kargin V. A., Papkov S. F. *Russ. J. Phys. Chem.* **1936**, *7*, 483–95.
4. Belousov V. P., Panov M. Yu. *Thermodynamics of Aqueous Solutions of Non-Electrolytes* Khimiya, Leningrad **1983**.
5. Krestov G. A. *Thermodynamics of Non-Electrolyte Solutions* IKhNR, Ivanovo **1989**, 3–7.
6. Nikiforov M. Yu., Alper G. A., Durov V. A. *Solutions of Non-Electrolytes in Liquids* Nauka, Moscow **1989**.
7. Krestov G. A., Berezin B. D. *Basic Notions of Modern Chemistry* Khimiya, Leningrad **1986**.
8. Krestov G. A. *Thermodynamics of Ionic Processes in Solutions* Khimiya, Leningrad **1984**.
9. Krestov G. A., Kolker A. M., Safonova L. P. *Dokl. Akad. Nauk SSSR* **1985**, *280*, 404–07.
10. Kobenin V. A., Kazanskii A. N., Krestov G. A. *Thermodynamic Properties of Solutions* IKhTI, Ivanovo **1984**, 3–19.
11. Krestov G. A. *Ionic Solvation* Nauka, Moscow **1987**, 5–34.
12. Solomonov B. I., Konovalov A. I. *Russ. J. Gen. Chem.* **1985**, *55*, 2529–46.
13. Abraham M. H. *J. Am. Chem. Soc.* **1979**, *101*, 5477–89.
14. Mourahamos J. J., Stien M. L., Reisse J. *Chem. Phys. Lett.* **1976**, *42*, 373–76.
15. Solomonov B. N., Borisover M. D., Konovalov A. I. *Russ. J. Gen. Chem.* **1986**, *56*, 1–12.
16. Solomonov B. N., Borisover M. D., Konovalova L. K., Konovalov A. I. *Russ. J. Gen. Chem.* **1986**, *56*, 1345–49.
17. Korolyov V. P., Vandyshev V. N., Krestov G. A. *Russ. J. Gen. Chem.* **1987**, *57*, 1813–17.
18. Saluja P. P. S., Peacock L. A., Fuchs R. *J. Am. Chem. Soc.* **1979**, *101*, 1958–62.
19. Gutmann W. *Chemistry of Coordination Compounds in Non-Aqueous Solutions* Mir, Moscow **1971**.
20. Arnett E. M., Jorris L., Mitchell E. *J. Am. Chem. Soc.* **1970**, *92*, 2365–77.
21. Arnett E. M., Murthy T. S. S. R., Schleyer P. V. R., Jorris L. *J. Am. Chem. Soc.* **1967**, *89*, 5955–57.
22. Duer W. C., Bertrand G. L. *J. Am. Chem. Soc.* **1970**, *92*, 2587–89.
23. Solomonov B. N., Konovalov A. I., Gorbachuk V. V. *Russ. J. Gen. Chem.* **1985**, *55*, 1889–1906.
24. Levina O. V., Iogansen A. V., Kurkchi G. A., Bayeva V. P. *Russ. J. Phys. Chem.* **1978**, *52*, 153–56.
25. Stephenson W. K., Fuchs R. *Can. J. Chem.* **1985**, *63*, 342–48.
26. Burger K. *Solvation, Ionic Reactions and Complex Formation in Non-Aqueous Media* Mir, Moscow **1984**.
27. Kamlet M. J. *Prog. Phys. Org. Chem.* **1981**, *13*, 485–630.
28. Gryzlov S. I., Golshtein I. P. in *Abstracts of IV All-Union Meeting on Problems of Solvation and Complex Formation in Solution* IKhNR, Ivanovo **1989**, *175*.
29. Balk R. W., Somsen G. *J. Phys. Chem.* **1985**, *89*, 5093–97.
30. Huyskens P. L., Huulait-Pirson M. C. *Organic Coating: Science and Technology Coatings Research Institute, Limelette, Belgium* **1986**, Vol. *8*, 155–73.
31. Bale C. W., Pelton A. D. *Met. Trans.* **1974**, *5*, 2323–37.
32. Myasoedova V. V., Krestov G. A. in *Abstracts of All-Union Conference on Problems of Complex Usage of Wood Raw Materials, Riga*, **1984**, 122–23.
33. Golova L. K., Andreyeva O. E., Kulichikhin V. G. *Vysokomol. Soyedin.* **1986**, *28*, 2308–12.
34. Novikova N. V., Matyushin Yu. N., Konkova T. S. *Izv. Akad. Nauk SSSR. Ser. Khim.* **1987**, *10*, 2319–21.
35. Isogai A., Ishizu A., Nakano J. *J. Appl. Polym. Sci.* **1987**, *33*, 1283–90.

36. Belousov V. P., Morachevskii A. G. *Heats of Mixing of Liquids* Khimiya, Leningrad **1970**.
37. Korolyov V. P. *Solutions of Non-Electrolytes in Liquids* Nauka, Moscow **1989**, p. 103–35.
38. McGlashan M. L. *Chemical Thermodynamics* Academic Press, London **1979**.
39. Belousov V. P., Morachevskii A. G., Panov M. Yu. *Thermal Properties of Non-Electrolyte Solutions* Khimiya, Leningrad **1981**.
40. Tager A. A. *Physicochemistry of Polymers* Khimiya, Moscow **1978**.
41. Meerson S. I. *Kolloid. Zh.* **1969**, *31*, 421–26.
42. Tsvetkov V. G. in *Investigation Methods for Cellulose* Karlivan V. P. (ed.) Zinatne, Riga **1981**, 126–37.
43. Prchal M., Dohnal V., Vesely F. *Collect. Czech. Chem. Commun.* **1982**, *47*, 3171–76.
43a. Brandreth D. A., O'Neill S. P., Missen R. W. *Trans. Faraday Soc.* **1966**, *62*, 2355.
43b. Wilson G. M. *J. Am. Chem. Soc.* **1964**, *86*, 127.
44. Nikolaenko I. V., Batalin G. I. *Teor. Eksp. Khim.* **1987**, *23*, 198–203.
45. Smirnova N. A. *Statistical Thermodynamics Methods in Physical Chemistry* Vysshaya Shkola, Moscow **1982**.
46. Ouken V., Gmehling J. *Chem. Ing. Tech.* **1977**, *49*, 404–11.
47. Fredenslund A., Jones Russel L., Prausnitz J. M. *AIChE J.* **1975**, *21*, 1086–99.
48. Smirnova N. A. *Molecular Theories of Solutions* Khimiya, Leningrad **1987**.
49. Van den Bery J. W. *Ind. Eng. Chem.* **1984**, *23*, 321–22.
50. Basedow A. M., Ebert K., Feigenbutz W. *Makromol. Chem.* **1980**, *181*, 1071–80.
51. Myasoedova V. V., Zavyalov N. A., Pokrovskii S. A., Krestov G. A. *Dokl. Akad. Nauk SSSR* **1988**, *303*, 901–04.
52. Basedow A. M., Ebert K. H., Emmert J. *Makromol. Chem.* **1979**, *180*, 1339–43.
53. Pokrovskii S. A. *PhD Thesis* Ivanovo **1989**.
54. Tsvetkov V. G., Kaimin I. F., Ioyelovich M. Ya., Prokhorov A. V. *Thermodynamics of Organic Compounds* GGU, Gorky **1982**, 54–60.
55. Genin A. L., Bakatskaya M. I., Nikitin I. V. *Vysokomol. Soyedin.* **1983**, *25*, 1717–22.
56. Abakshin V. A., Krestov G. A. *Dokl. Akad. Nauk SSSR.* **1986**, *291*, 1135–37.
57. Keizo M. *Macromolecules* **1984**, *17*, 449–52.
58. Balk R. W., Somsen G. *J. Chem. Soc., Faraday Trans. 1* **1986**, *82*, 933–42.
59. Tsvetkov V. G. *Thermodynamics of Organic Compounds* GGU, Gorky **1986**, 85–93.
60. Golova L. K., Kulichikhin V. G., Papkov S. P. *Vysokomol. Soyedin.* **1986**, *28*, 1795–1809.
61. Novikova N. V., Matyushin Yu. N., Konkova T. S. *Izv. Akad. Nauk SSSR, Ser. Khim.* **1987**, 2319–21.
62. Novikova N. V., Matyushin Yu. N., Sopin V. P. *Izv. Akad. Nauk SSSR, Ser. Khim.* **1987**, 2095–98.
63. Matyushin Yu. N., Vorobyov A. V., Novikova N. V. in *Abstracts of Scientific Seminar on Investigation Methods for Cellulose, Riga* **1988**, 139–42.
64. Tsvetkov V. G. *DSc Thesis* Gorky **1986**.
65. Usmanov Kh. U., Khakimov I. Kh. *Uzb. Khim. Zh.* **1959**, *2*, 21–28.
66. Kuvshinnikova S. A., Bernardelli A. E., Mishchenko K. P. *Papers of Leningrad Institute of the Cotton Industry* **1970**, *25*, 98–105.
67. Marchenko G. N., Tsvetkov V. G., Marsheva V. N. *Thermodynamic Properties of Solutions* IKhTI, Ivanovo **1984**, 37–39.
68. Zenkov I. D., Zalenukhin R. V., Solovetskii V. 1., Papkov S. P. *Vysokomol. Soyedin.* **1988**, *30*, 1718–22.
69. Myasoedova V. V., Zavyalov N. A., Pokrovsky S. A., Krestov G. A. *Thermochim. Acta* **1990**, *169*, 111–19.

70. Tager A. A. *Vysokomol. Soyedin.* **1984**, *25*, 659–74.
71. Rabek J. *Experimental Methods in Polymer Chemistry, Part 1* Mir, Moscow **1983**.
72. Bickles N., Segal L. (eds) *Cellulose and Its Derivatives* Mir, Moscow **1974**.
73. Lard J., Avedikian L., Perron G. *J. Solution Chem.* **1981**, *10*, 3100–20.
74. Vandyshev V. N., Korolyov V. P., Krestov G. A. in *Abstracts of V All-Union Conference on the Thermodyamics of Organic Compounds, Kuibyshev* **1987**, 183.
75. Meerson S. I., Petryuk A. M., Vasserman A. M. *Cellul. Chem. Technol.* **1977**, *11*, 173–89.
76. Pokrovsky S. A., Zavyalov N. A., Myasoedova V. V. *Dep. VINITI, N7, 74-B.*
77. Myasoedova V. V., Alexeyeva O. V., Krestov G. A. *Russ. J. Appl. Chem.* **1987**, *60*, 2523–26.
78. Myasoedova V. V., Adamova O. A., Krestov G. A., *Vysokomol. Soyedin. Krat. Soobshch.* **1984**, *26*, 215–17.
79. Myasoedova V. V., Zavyalov N. A., Pokrovsky S. A., Krestov G. A. in *International Conference on Chemical Thermodynamics and Calorimetry Beijing*, **1989**, Progress Booklet A42.
80. Daragan V. A., Ilyina E. E., Myasoedova V. V., Safronkin P. G. *Izv. Akad. Nauk SSSR, Ser. Khim.* **1989**, *8*, 1766–68.
81. Safronov A. P., Tager A. A., Voit V. V. *Vysokomol. Soyedin.* **1988**, *30*, 2360–64.
82. Marcus Y. *J. Solution Chem.* **1984**, *13*, 599–624.
83. Pokrovsky S. A., Zavyalov N. A., Telegin F. Yu., Myasoedova V. V. in *Abstracts of VI All-Union Conference on Thermodynamics of Organic Compounds, Minsk*, **1990**, 231.
84. Lvovskii E. N. *Statistical Methods of Building Empirical Formulae* 2nd ed. Vyssh. Shk., Moscow **1988**, 46–69, 68–73.
85. Zavyalov N. A., Myasoedova V. V., Krestov G. A. in *Abstracts of VI All-Union Conference on Thermodynamics of Organic Compounds, Minsk* **1990**, 102.
86. Zavyalov N. A., Pokrovsky S. A., Yuryev I. K., Myasoedova V. V. in *Abstracts of II All-Union Conference on Chemistry and Application of Non-Aqueous Solutions, Kharkov* **1989**, 105.
87. Tarchevskii I. A., Marchenko G. N. *Biosynthesis and Structure of Cellulose* Nauka, Moscow **1985**.
88. Zhbankov R. P., Kozlov P. V. *Physics of Cellulose and Its Derivatives* Nauka i Tekhnika, Minsk **1983**.
89. Rogovin Z. A., Galbraikh L. S. *Chemical Transformations and Modification of Cellulose* Khimiya, Moscow **1979**.
90. Rogovin Z. A., Shorygina N. N. *Chemistry of Cellulose and Its Sputniks.* Goskhimizdat, Moscow **1953**.
91. Sarkanen K. V. in *Chemistry of Wood I* Brauning B. P. (ed.) Lesnaya Promyshlennost, Moscow **1967**, 184.
92. Rogovin Z. A. *Basis of Chemistry and Technology of Production of Chemical Fibres*, Vol. 1, Khimiya, Moscow **1964**.
93. Papkov S. P., Fainberger E. S. *Interaction of Cellulose and Cellulose Materials with Water* Khimiya, Moscow **1976**.
94. Klenkova N. I. *Structure and Reactivity of Cellulose* Nauka, Leningrad **1976**.
95. Zharkovskii D. V. *Physico-Chemical Investigations of Cellulose and Its Derivatives* Nauka i Tekhnika, Minsk **1960**.
96. Matyushin Yu. N., Kurchatova L. I., Sopin V. F. in *Abstracts of I All-Union Conference on Cellulose Synthesis and Its Regulation, Kazan*, **1980**, 36.
97. Rabek J. *Experimental Methods in Polymer Chemistry, Part 2* Mir, Moscow **1983**.
98. Zharkovskii D. V., Emelyanova S. M. *Izv. Akad Nauk BSSR* **1964**, *1*, 61–64.
99. Filipp B. *Vysokomol. Soyedin.* **1981**, *23*, 3–13.

100. Nikitin V. M., Obolenskaya A. V., Shchegolev V. P. *Chemistry of Wood and Cellulose* Lesnaya Promyshlennost, Moscow **1978**.
101. Dumanskii A. V., Nekryach E. F. *Nauchn. Tr. Leningr. Gos. Univ.* **1960**, *91*, 3–10.
102. Zorina R. I., Mikramilov Sh. M., Shapovalov O. I. *Khim. Dreves.*, **1983**, *2*, 3–6.
103. Kiselev V. P., Shakhova G. M., Fainberg E. Z. *Vysokomol. Soyedin.* **1976**, *18*, 847–49.
104. Nikitin I. V., Vainshtein E. F., Kushnerev M. Ya. *Dokl, Akad, Nauk SSSR* **1982**, *262*, 900–04.
105. Mikhailov N. V., Fainberg E. Z. *Dokl. Akad. Nauk SSSR* **1956**, *109*, 109–14.
106. Mikhailov N. V., Fainberg E. Z. *Dokl. Akad. Nauk SSSR* **1956**, *109*, 1160–62.
107. Fainberg E. Z., Mikhailov N. V., Papkov S. P. *Vysokomol. Soyedin.* **1967**, *9*, 1483–87.
108. Tsvetkov V. G., Rabinovich I. B., Kaimin I. F. in *Abstracts of All-Union Conference on Physical and Physical-Chemical Aspects of Cellulose Activation, Riga*, **1981**, 110–13.
109. Dole B. E., Tsao G. *J. Appl. Polym. Sci.* **1982**, *27*, 1233–41.
110. Ioyelovich M. Ya. *Khim. Dreves.* **1985**, *5*, 111–12.
111. Jones M. (ed.) *Biochemical Thermodynamics* Mir, Moscow **1982**.
112. Tager A. A., Kargin V. A. *Russ. J. Phys. Chem.* **1941**, *15*, 1036–54.
113. Tager A. A., Popova O. V. *Russ. J. Phys. Chem.* **1959**, *33*, 593–98.
114. Zavyalov N. A., Pokrovsky S. A., Myasoedova V. V., Krestov G. A. *Thermodynamics of Non-Electrolyte Solutions, Collected Papers of IKhNR* IKhNR Ivanovo **1989**, 60–66.
115. Sokolov V. V., Ivanov A. V., Poltoratskii G. M. in *Abstracts of IX All-Union Conference on Calorimetry and Chemical Thermodynamics, Tbilisi* **1982**, 181.
116. Ivanov A. V., Tsvetkov V. G., Sokolov V. V. in *Abstracts of V Republican Conference of Young Scientists–Chemists, Tallin* **1983**, 201.
117. Zakharova L. N., Novikov V. B., Sopin V. F. in *Abstracts of V All-Union Conference on Thermodynamics of Organic Compounds, Kuibyshev* **1987**, 75.
118. Makitra R. G., Prig Ya. N., Kavelyuk R. B. *The Most Important Characteristics of Solvents Used in LSE Equations* LPI, L'vov **1986**.
119. *A Chemist's Reference-Book*, Vol.1 GKhI, Moscow **1963**.
120. Karapetyan Yu. A., Eichis V. N. *Physico-Chemical Properties of Electrolyte Non-Aqueous Solutions* Khimiya, Moscow **1989**.

CHAPTER 4

Solvation of Cellulose and Its Derivatives in Non-aqueous Solutions

High-resolution NMR spectroscopy has been used in recent years to investigate cellulose [1]. Cross-polarization magic angle spinning ^{13}C NMR has given much information about the solid phase. Nehls *et al.* [1] reviewed the use of ^{13}C NMR to investigate cellulose and its derivatives in the liquid phase. They discussed the advantages of the use of ^{13}C rather than ^{1}H NMR spectroscopy for investigating the dissolution processes which occur when cellulose is dissolved in non-derivatizing and in derivatizing solvents and for the characterization of soluble cellulose derivatives. In principle, the two techniques should yield equivalent information but in practice the ^{1}H NMR spectra of polymers consist of broad overlapping peaks owing to enhanced dipolar interaction resulting from hindered isotropic motion within the molecule. The effect is much less in the case of ^{13}C NMR and, in addition, the chemical shift is greater. As a result, much more information can be obtained from ^{13}C NMR spectra.

The use of the spin probe method to study dynamics and interactions in polymer solutions is well established. However, it is only recently that the application of this technique to cellulose solutions has yielded significant results. This has been due to the difficulties in the selection of systems with high concentrations of cellulose or its derivatives and also to the difficulties in the selection of suitable compounds to use as spin probes. Studies of H-complexes in non-aqueous solutions of low-molecular-mass systems and also of model systems are a help in the understanding of problems concerning partially or completely substituted derivatives of cellulose. It has been found that solutions of sugars, such as cellobiose or methyl-α-D-glucoside, dissolved in dimethyl sulfoxide (DMSO-d_6), have values of the chemical shifts for the protons of hydroxy groups which do not agree with predictions of electronic theory. Efforts have been made to explain this discrepancy.

1 THE USE OF NUCLEAR MAGNETIC RELAXATION OF MOLECULAR PROBES TO ESTIMATE THE FORMATION OF H-COMPLEXES IN NON-AQUEOUS SOLUTIONS OF CELLULOSE DERIVATIVES

This chapter deals with the investigation of solutions of cellulose and its derivatives by molecular NMR relaxation probes (MRPs). In this technique, a proportion of small molecules are added to a polymer solution. The motions of these small molecules are investigated by magnetic relaxation of nuclei. Active MRPs are molecules that form complexes with polymers (e.g. by means of hydrogen bonding), competing successfully with the solvent. Inert MRPs have a lower ability than solvent molecules to form complexes with the polymer.

Cellulose and its derivatives are very convenient for such investigations, since all large groups in an elementary link of a chain are in the equatorial position. As a result, the steric factors oppose any changes to the normal 'chair' configuration, C_1. Reliable values of the chemical shifts for a series of cellulose acetates in the same solvent can be obtained.

Cellulose derivatives with various degrees of substitution, dissolved in completely deuterated DMSO, have been studied [2–5]. 1,4-Diazobicyclo [2.2.2] octane (DABO) was the active probe. This compound has a strong tendency to form hydrogen bonds with non-substituted OH groups of cellulose derivatives. It has also has a simple ^{13}C NMR spectrum, which makes relaxation measurements easy. The introduction of small (3–5 mass%) additions of DABO increases the solubility of cellulose acetates in DMSO, which permits the measurements to be carried out at high polymer concentrations. Tetralin served as the inert MRP. The magnetic relaxation rate of ^{13}C nuclei in positions 2 and 3 served as the indicator of tetralin rotary mobility. The signals from these nuclei have maximum relaxation rates. This allowed the corresponding relaxation curves to be readily obtained.

This method can be used to determine the degree of substitution of hydroxyl by acetate groups in cellulose macromolecules. The author and her colleagues have measured the spin–lattice relaxation rates of ^{13}C nuclei, W, of DABO and tetralin at fixed concentrations of polymer with various degrees of substitution DS [5–7]. For tetralin, at different magnitudes of DS, the relaxation rates, W, did not change appreciably.

In the case of DABO, there was a linear dependence of W on DS. This dependence is shown in Figure 4.1. The large change in the case of DABO suggests that monitoring of W could be used for analytical purposes.

The dimensions of DABO and tetralin molecules are similar. Differences in the relaxation rates of ^{13}C nuclei, and hence differences in rotary mobility, can only be due to complex formation between DABO molecules and free OH groups of the polymer.

The variation of W of DABO and tetralin with the concentration, C, of cellulose acetates with different values of DS have been measured [3], as shown in Figures 4.2

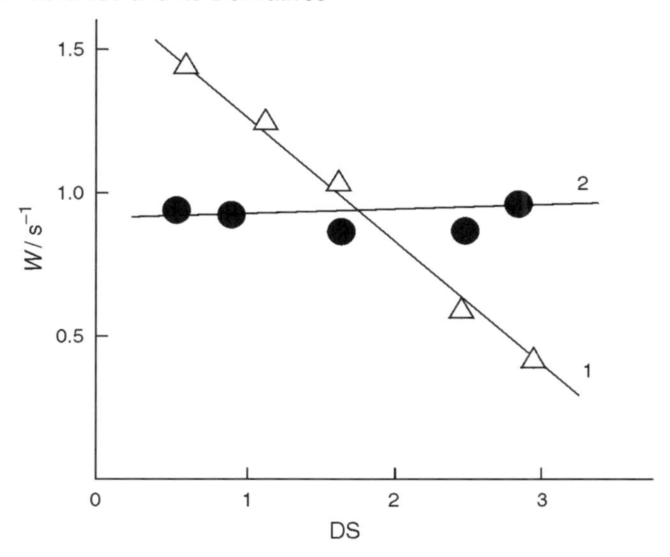

Figure 4.1 Dependence of the relaxation rate, *W*, on the degree of substitution of cellulose acetates in DMSO at polymer concentration 300 mg cm^{-3}. 1, DABO; 2, tetralin

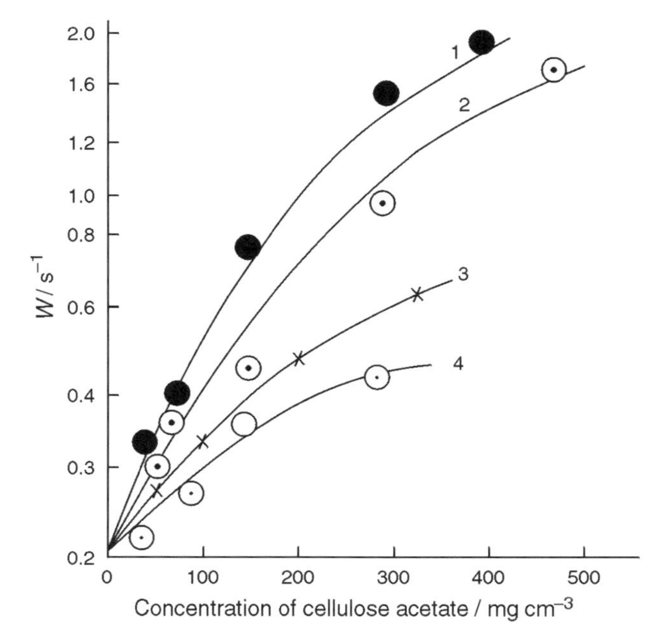

Figure 4.2 Dependence of the relaxation rate, *W*, of DABO on the concentration of cellulose acetates with different degrees of substitution: 1, DS = 0.60; 2, DS = 1.61; 3, DS = 2.38; 4, DS = 2.90

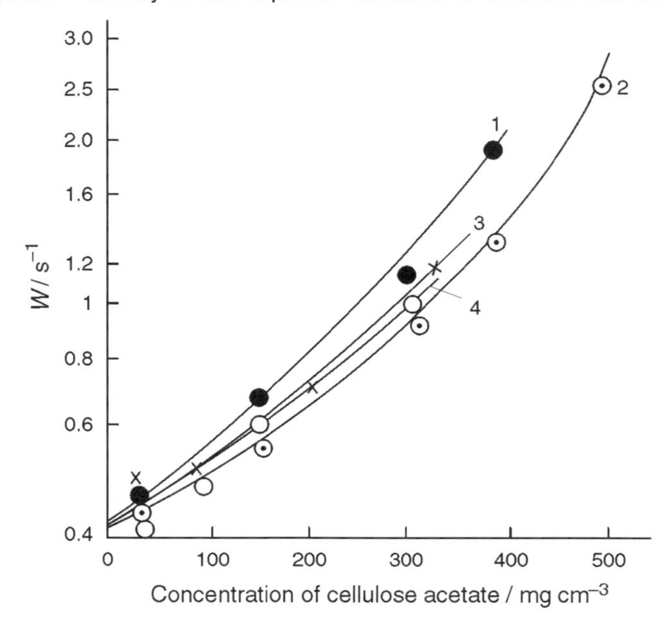

Figure 4.3 Dependence of the relaxation rate of tetralin on the concentration of cellulose acetates with different degrees of substitution: 1, DS = 0.60; 2, DS = 1.66; 3, DS = 2.38; 4, DS = 2.90

and 4.3. It was found that the variations of W with C were approximated by equations of the form

$$W = W_0 \exp\left(C/C_0 + aC^2\right) \tag{4.1}$$

Where C_0 and a are constants for the particular solvent and degree of substitution. In the case of tetralin the value of a is zero.

The increase in tetralin relaxation rate with increasing polymer concentration is due to restriction of the rotary mobility by steric factors and by the increase in the ordering of surrounding molecules [4].

There is a simple explanation for the concentration dependence of DABO relaxation rates. It has been shown [6] that

$$W = XW_b + (1 - X)W_f \tag{4.2}$$

where W is the experimentally observed value of the relaxation rate, W_b and W_f are the relaxation rates of bonded and free DABO molecules, respectively, and X is the fraction of free DABO molecules. The simplest scheme of complex formation of DABO with OH groups of polymer is as follows:

$$\mathrm{DABO} + \mathrm{OH} \overset{k}{\rightleftharpoons} \mathrm{DABO} \cdots \mathrm{OH} \tag{4.3}$$

Assuming that

$$[\mathrm{DABO}\cdots\mathrm{OH}] \ll C_{\mathrm{DABO}} + C_{\mathrm{OH}} \tag{4.4}$$

where C_{DABO} and C_{OH} are the total concentrations of molecules of DABO and of OH groups (free + participating in complex formation in each case), it can be shown that

$$W = \frac{1}{1 + K[C_{\mathrm{DABO}} + C_{\mathrm{OH}}]}[W_{\mathrm{b}}KC_{\mathrm{OH}} + W_{\mathrm{f}}(1 + KC_{\mathrm{DABO}})] \tag{4.5}$$

Equation (4.5) is correct only at low polymer concentrations. With increasing concentration of the cellulose acetates, the number of OH groups that are accessible to the probe begins to depend strongly on concentration. This is because accessibility is determined by interaction between macromolecules. At high concentrations the C_{OH} values in Eq. (4.5) should be modified to reflect the change in the relative number of polymer OH groups participating in the interaction with DABO molecules. For the qualitative description of this process a Gaussian equation can be used:

$$C'_{\mathrm{OH}} = C_{\mathrm{OH}} \exp(-C^2_{\mathrm{OH}}/\bar{C}^2_{\mathrm{OH}}) \tag{4.6}$$

where C'_{OH} is a certain characteristic concentration of OH groups reflecting the interaction between macromolecules in solution and $\bar{C}_{\mathrm{OH}}$ is the mean concentration of OH groups. If it is assumed that the accessibility of OH groups for the probe molecules changes to a small extent with the degree of substitution, then the following correlation below between $\bar{C}_{\mathrm{OH}}$ and DS will exist:

$$\bar{C}_{\mathrm{OH}} = \bar{N}_{\mathrm{link}}(3 - \mathrm{DS}) \tag{4.7}$$

where $\bar{N}_{\mathrm{link}}$ is the average concentration of the number of links of the polymer chain per unit volume of the solution.

It should be noted that relaxation rates W_{f} and W_{b} are also functions of polymer concentration. The form of concentration dependence, $W_{\mathrm{f}} = f(C_{\mathrm{OH}})$, for DABO should be similar to the dependence for tetralin due to the closeness of the dimensions of these molecules, i.e.

$$W_{\mathrm{f}}(C_{\mathrm{OH}}) = W_{\mathrm{f}}(C_{\mathrm{OH}} = 0)\frac{W_{\mathrm{T}}(C_{\mathrm{OH}})}{W_{\mathrm{T}}(0)} \tag{4.8}$$

where W_{T} is the relaxation rate for tetralin.

The dependence of W_{b} on C_{OH} is more complex, since it is mainly determined by the dynamics of the polymer chain. It can be described in the general case by the relationship

$$W_b(C_{OH}) = W_b(C_{OH} \to 0)(1 + \alpha C_{OH}^{\beta}) \tag{4.9}$$

where α and β are parameters with $\beta = 1.5$ [4]. A reduced relaxation rate, W^*, can be defined as $W^* = W/W_f$. If complex formation is absent, then $W^* = 1$ over the entire interval of polymer concentrations. Calculations of $W^*(C_{OH})$ have been carried out using Eqs (4.5)–(4.9) for various values of the parameters in these equations [6].

The variation of W^* with C_{OH} passes through a maximum. The height of this depends on many parameters, but its position is determined by the value of C_{OH} only. Hence the position of the maximum is a measure of the accessibility of OH groups in a polymer. It yields information on macromolecular structure and on supramolecular structure.

The variation of W^* of DABO with the magnitude of $\bar{N}_{link}$ was calculated from experimental data and is plotted in Figure 4.4.

The position of the maxima of W^* moves to higher values of $\bar{N}_{link}$ with decreasing degree of substitution, DS. This shows that this parameter is not constant for the polymers investigated. At high DS values there is a more rapid 'screening' of free OH groups of the cellulose acetates than at low values. This may indicate that the formation of hydrogen bonds between hydroxyl and acetate groups is more

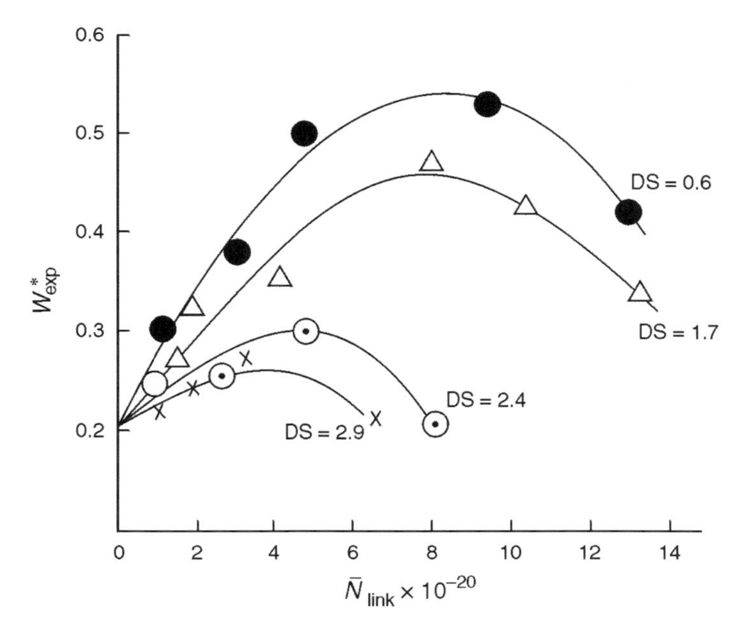

Figure 4.4 Dependence of the reduced relaxation rate of DABO, W^*_{exp}, on the number of links, $\bar{N}_{link}$, in cellulose acetates with different degrees of substitution, DS

likely to occur than the formation of bonds between the hydroxyl groups of one or different polymer molecules. The high electronegativity and acceptor ability of the acetate groups and their higher conformational mobility all play an important role in this process [5].

Ilyina *et al.* [5] have reported that spin–lattice relaxation rates and self-diffusion coefficients of active molecular probes depend on the degree of substitution of cellulose acetate. Information on the hydroxyl groups in cellulose can be obtained by analysing the characteristics of the rotational and translational motion of the probes when the concentration of polymer is changed. The binding capacity of the hydroxyl groups was found to be higher and their shielding less the greater was the degree of substitution by acetate groups. The polymer network was apparently thicker the lower was the degree of substitution.

It is interesting to compare these results with the increase in Kuhn segment for cellulose acetates with increasing DS (Chapter 2). There is an interrelation between the rigidity of the polymer chain and the change in the character of intra- and intermolecular interactions. This results in change in the screening of the reactive centres of the chain.

Polymer–polymer interactions have been studied in solutions of mixtures of CA with polyethylene glycols (PEG), $HO(CH_2CH_2O)_nH$, dissolved in non-aqueous solvents [8]. Spin–lattice relaxation rates of ^{13}C nuclei of PEG molecules in CA solutions of different concentrations of acetone, dimethylformamide (DMF) and DMSO were measured. Measurements have shown that the way in which the rotary mobility of PEG changes with changes in concentration of CA depends on the donor–acceptor properties of the solvent. In the least electron-donating solvent, acetone (Figure 4.5), there occurs the effect which the author called 'reversibility of mobility:' with increasing CA concentration. The rotary mobility of PEG with mass 400 is gradually overtaken by the mobility of PEG 15 000. At high CA concentrations the spin–lattice relaxation rate of PEG 15 000 is the greater.

Complementary binding of the oxygens of PEG with hydroxyl groups of CA is a possible reason for the effect. Here, owing to the thermal fluctuations, the length of the section of bonding is limited. PEG molecules with greater molecular mass remain largely non-bonded. This results in an increase in the average mobility of these molecules relative to PEG molecules with small mass, which are bonded almost completely. In the case of the strongest of the electron-donating solvents investigated, DMSO, the effect of reversibility of mobility is not observed (Figure 4.5); this may be due to PEG competing weakly with DMSO molecules in the formation of hydrogen bonds with residual CA groups. In DMF, with intermediate donor properties, the measurements were carried out with samples of CA having different values of DS: 1.66 and 2.38 (Figure 4.6). It was found that in CA samples with DS = 2.38, the mobilities of PEG with different masses change in a similar fashion with increasing CA concentration. When DS = 1.66, the values of W of PEG 400 and PEG 15 000 become closer with increasing concentration, C_{CA}. There is no reversibility of mobility [7].

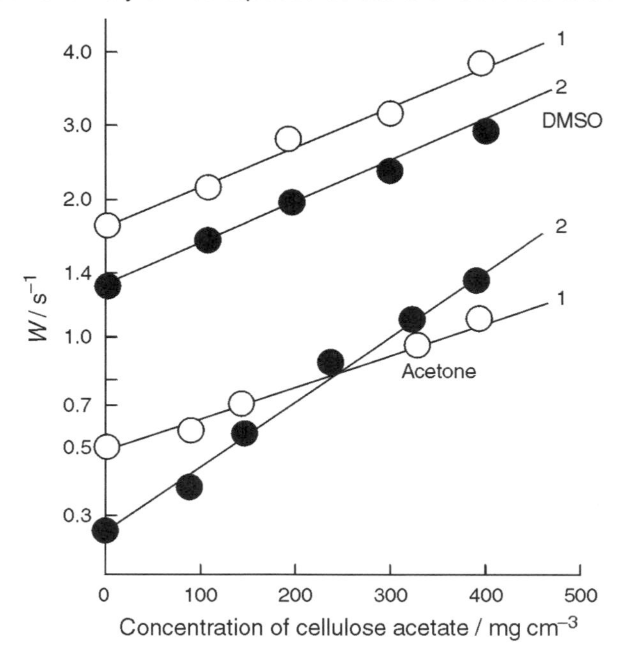

Figure 4.5 Variation of relaxation rate of PEG, *W*, with concentration of cellulose acetate dissolved in DMSO and in acetone: 1, PEG ($M_n = 400$); 2, PEG ($M_n = 15\,000$)

The hypothesis that there is complementary binding of PEG oxygen atoms with free OH groups of CA is supported by the variation of relaxation rates of ^{13}C nuclei in PEG with relative molecular mass, M_n, of PEG (Figure 4.7). *W* would not depend on M_n if there were no binding between PEG and CA. In acetone, when the polymer chains are short and M_n is low, *W* has a high value. The proportion of the length of a typical PEG chain which is bonded to CA molecules is high under these circumstances. With increasing M_n, *W* decreases owing to the decreasing proportion of the length of a chain which is bonded to CA molecules. *W* does not vary appreciably with M_n in DMSO. This is additional evidence that there is strong competition with DMSO molecules for the formation of hydrogen bonds with OH groups of CA.

The solvent can also influence the state of PEG molecules in solutions. The reversibility of PEG mobility in acetone solutions has been discussed above. The way in which PEG relaxation rates, *W*, change with M_n in acetone solution differs from that in DMSO solution. This can be explained by the tangles becoming more compact in the latter solvent. The more compact the tangles, the greater is the effect on *W* of their motion as a whole. An increase in M_n leads to larger and more slowly moving tangles and hence larger values of *W*. In addition, the larger and more compact the tangles, the more the rotations of the methylene groups are inhibited

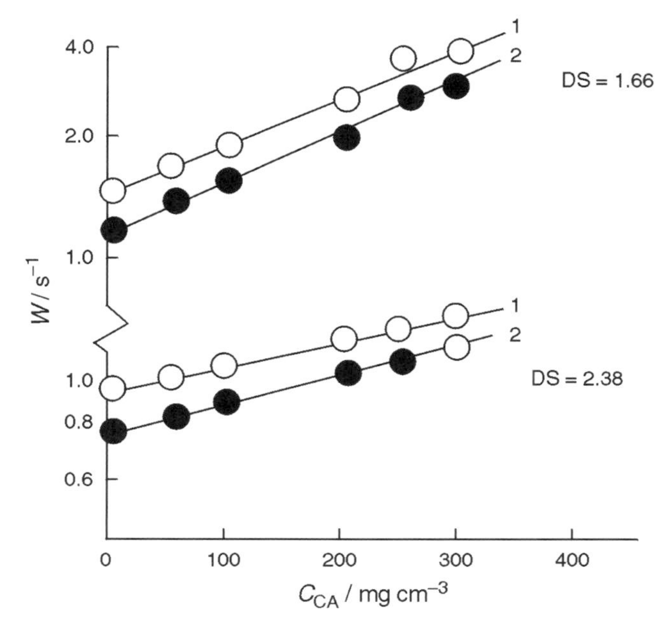

Figure 4.6 Variation of relaxation rate of PEG, W, with concentration of cellulose acetate dissolved in DMF. DS is the degree of substitution of the cellulose acetate. 1, PEG ($M_n = 400$); 2, PEG ($M_n = 15\,000$)

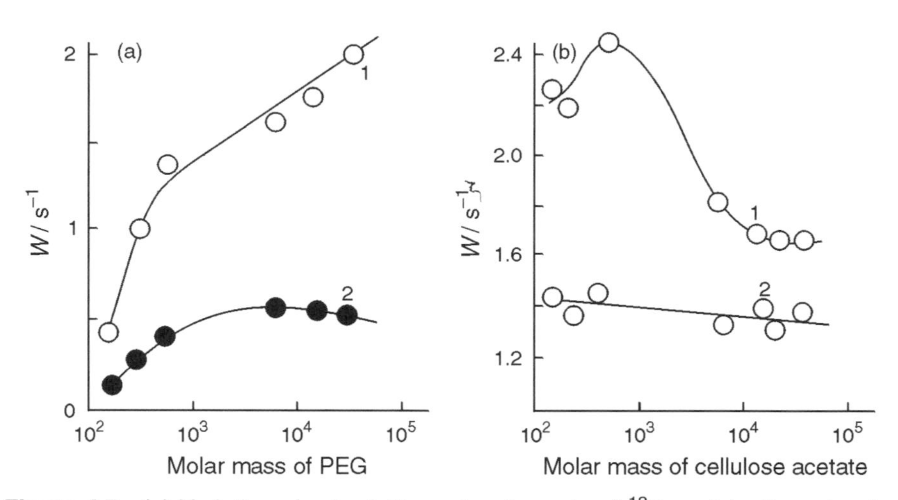

Figure 4.7 (a) Variation of spin–lattice relaxation rate of ^{13}C nuclei with molecular mass of PEG in solutions of cellulose acetates (CA): 1, in acetone; 2, in dimethyl sulfoxide. (b) Variation of spin–lattice relaxation rate of ^{13}C nuclei with molecular mass of CA: 1, in acetone; 2, in dimethyl sulfoxide

and the greater is the value of *W*. In the case of acetone the dependence of *W* on M_n reaches a plateau. At this point the segmental mobility of PEG in acetone does not change with increase in M_n because of the loose structure of the tangles in this solvent.

The translatory motion of molecular probes characterized by values of the coefficient of translatory diffusion, *D*, can give information about the structural and chemical nature of polymer solutions. Coefficients of translatory diffusion of molecular probes have been measured at 25 °C by Daragan *et al.* [8], who used a spin echo method with a pulse gradient of magnetic field [9, 10] in a modernized IVR-1 pulse relaxometer [8].

Measurements of *D* of tetralin and DABO in solutions of cellulose acetates at a concentration at 250 mg cm^{-3} and various values of DS were carried out to check the sensitivity of the measurements to the chemical composition of polymers. DABO and tetralin molecules have similar dimensions. The variation of *D* with DS for DABO (Figure 4.8) is much greater than for tetralin. In both cases the values of *D* increase linearly with values of DS. The change for DABO is the more marked. This is likely to be related to complex formation of DABO with free OH groups in cellulose acetate. Tetralin does not take part in similar complex formation. The

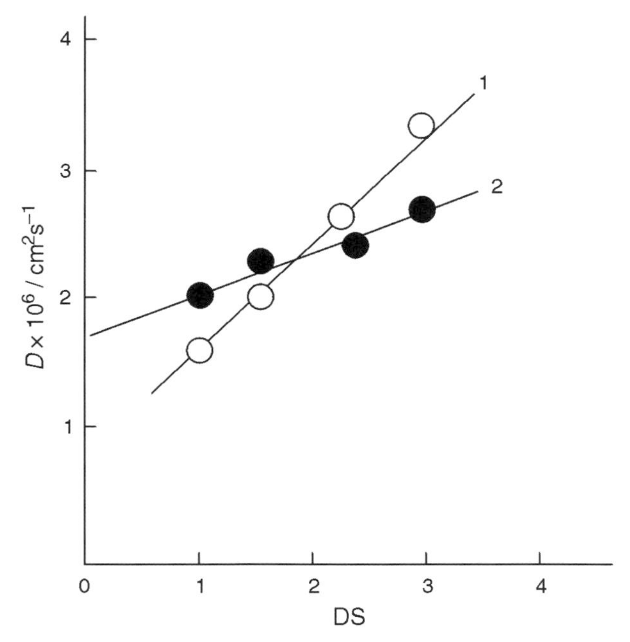

Figure 4.8 Dependence of the coefficient of translatory diffusion (*D*) of ^{13}C nuclei of molecular probes on the degree of substitution of cellulose acetates in DMSO solutions (300 g dm^{-3}). 1, DABO; 2, tetralin

measurement of D for DABO could be used as a rapid method of measuring DS for cellulose acetate.

The effect of change of DS on the value of D for tetralin in these solutions is in marked contrast to the effect on the rotary mobility, which is negligible. The supramolecular structure of a polymer solution influences the translatory motion of inert MRPs more than their rotary motion. This follows when measuring D, since the displacement of MRP is registered at a distance which is noticeably greater than the dimensions of one macromolecule.

The change in values of D of tetralin can be explained by the difference in the structure of the spatial nets that are formed by the cellulose acetate macromolecules with different degrees of substitution. An increase in the value of DS means fewer free hydroxyl groups, fewer hydrogen bonds between cellulose acetate molecules, a looser spatial net, greater freedom of movement for tetralin molecules and a greater value of D. When DABO is present an increase in DS has the additional effect of reducing hydrogen bonding between DABO and cellulose acetate.

Masson and St John Manley [11] made solid-state NMR relaxation measurements on blended films of cellulose with poly(vinyl alcohol), polyacrylonitrile, poly(ε-caprolactone) and nylon-6. They were able to estimate the scale of miscibility of the different samples. Their results suggested that polyesters and nylons with a higher number of interacting groups per methylene unit would show better miscibility with cellulose than poly(ε-caprolactone) or nylon-6.

Walderhaug *et al.* [12] used pulsed field gradient NMR to investigate the self-diffusion of surfactants in gelling and non-gelling aqueous systems on ethyl(hydroxyethyl) cellulose (EHEC) to which had been added either sodium dodecyl sulfate (SDS) or cetyltrimethylammonium bromide (CTAB). Interactions between polymer and surfactant were stronger in the EHEC–SDS system than in the EHEC–CTAB system.

Nunes *et al.* [13] used ^{13}C NMR with cross-polarization and magic angle spinning (CP/MAS) to investigate the properties of cellulose acetate and cellulose propionate, both in the form of powder and in different membranes. They showed that the shielding asymmetry factor, measured on samples with a regular shape and with a fixed orientation in the rotor, can give information about ordered domains in the polymers.

Masson and St John Manley [14] used ^{13}C CP/MAS NMR to investigate blends of methylolcellulose and poly(4-vinylpyridine) and to make comparisons with similar blends of cellulose and poly(4-vinylpyridine). They reported that methylolcellulose is soluble in several solvents in which cellulose itself is insoluble. The preparation of cellulose blends via the formation of the methylol derivative is therefore a method of extending the range of synthetic polymers which can be blended.

Nehls *et al.* [1] have shown how ^{13}C NMR spectroscopy can be used to investigate regioselectivity in the formation of cellulose derivatives.

2 IR SPECTROSCOPY OF NON-AQUEOUS SOLUTIONS OF CELLULOSE

Measurements have been made of the IR spectra of solutions of cellulose in *N*-methylmorpholine-*N*-oxide, in 3-methylpyridine-*N*-oxide and in their mixtures. There is an increase in the integral intensity and in the half-width and also a displacement towards lower wavenumbers of the band due to the stretching vibration of the NO group. This indicates an interaction between NO and HO groups leading to a destruction of intermolecular hydrogen bonds between cellulose molecules [15, 16].

REFERENCES

1. Nehls I., Wagenknecht W., Philipp B., Stscherbina D. *Prog. Polym. Sci.* **1994**, *19*, 28–78.
2. Daragan V. A., Ilyina E. E., Myasoedova V. V., Prizment A. L. *Abstracts of IV All-Union Meeting on Problems of Solvation and Complex Formation in Solution, Ivanovo; IKhNR AN SSSR* **1989**, *June*, 138.
3. Daragan V. A., Ilyina E. E., Kuznetsova N. Ya., Myasoedova V. V. *Izv. Akad. Nauk SSSR, Ser. Khim.* **1989**, *8*, 1766–68.
4. Daragan V. A., Ilyina E. E., Myasoedova V. V. *Acid–Base Interactions and Solvation in Non-Aqueous Media* KhGU, Kharkov **1987**, 197.
5. Ilyina E. E., Daragan V. A., Prisment A. E. *Macromolecules* **1993**, *26*, 3319–23.
6. Daragan V. A. *DSc, Thesis* Moscow **1987**.
7. Daragan V. A., Safronkin P. G., Ilyina E. E., Myasoedova V. V. *Thermodynamics of Organic Compounds* KPI, Kuibyshev **1987**, 182.
8. Myasoedova V. V., Marchenko C. N., Krestov G. A. *Non-Aqueous Solution of Cellulose and its Derivatives* Nauka, Moscow **1991**.
9. Steiskal E. O., Ganner J. E. *J. Chem. Phys.* **1965**, *42*, 288–91.
10. Maklakov A. I., Skirda V. D., Fatkulin N. F. *Self-Diffusion in Solutions and Melts of Polymers* KGU, Kazan **1987**.
11. Masson J.-F., St John Manley R. *Macromolecules* **1992**, *25*, 589–92.
12. Walderhaug H., Nyström B., Hansen F. K., Lindman B. *J. Phys. Chem.* **1995**, *99*, 4672–78.
13. Nunes T., Burrows H. D., Bastos M., Feio G., Gil M. H. *Polymer* **1995**, *36*, 479–85.
14. Masson J.-F., St John Manley R. *Macromolecules* **1991**, *24*, 5914–21.
15. Kulichikhin V. G., Golova L. K. *Khimi. Dreves.* **1986**, 9–27.
16. Myasoedova V. V., Krestov G. A. *Papers of II All-Union Conference on Liquid-Crystalline Polymers, Vladimir* **1987**, 36.

CHAPTER 5

Mathematical Models of Cellulose and Its Derivatives in Solution

The application of simulation methods is one of the most important developments in the investigation of the physical chemistry of polymers. The selection of the corresponding models, their analysis and optimization are carried out using computers. The development of computers has resulted in extensive simulation of various physical systems. In particular, these techniques have been applied to numerous disordered systems with a strong interparticle interaction. These systems include molecular liquids, liquid metals, dense gases and plasma, solutions and melts of polymers and liquid crystals. Until recently, it was considered that cellulose did not form real solutions. As a consequence, all experimental evidence for the conformational characteristics of its molecules in the liquid phase was obtained from investigations on solutions of cellulose derivatives in particular esters. There is great interest in improving calculation techniques so that parameters characterizing cellulose itself in solution can be determined with greater accuracy. Despite the difficulties due to the complexity of the problem, this seems to the only way of obtaining information on the conformation of cellulose in solution. Direct experimental measurement cannot, as yet, give the information. Some calculations of this type have been carried out on several linear and cyclic macromolecules [1–13]. These studies give an insight into the problems of studying cellulose and its derivatives.

There are three main computer simulation methods which have been used, the Monte Carlo (MC), molecular dynamics (MD) and Brownian dynamics (BD) methods. In the MC method, the properties required are obtained by averaging over the ensemble of the configurations of the system that are generated by the computer by using an appropriate algorithm. The MD method gives the most detailed information of both equilibrium and dynamic properties. This method involves the numerical integration of the equations of motion of classical mechanics for all the particles of a system under given initial conditions. The BD method results in a rougher picture. In this method, the influence of the medium (solvent) is taken into account in terms of the action of the forces of viscous friction and of the action of random forces.

Most of the papers on the computer simulation of polymers by the MC method, following the first paper by Wall *et al.* [1], have dealt with an investigation of the influence of long-range interactions in a specified volume on the conformation of an isolated macromolecule. The macromolecule in such calculations was usually represented as a broken line wandering irregularly within a rectilinear lattice of a given type (cubic, tetrahedral, etc.). The surrounding solvent was not taken into account in explicit form, supposing that its role only involved some effective resetting of the interaction potentials between the macromolecule links. Early papers of this type have been reviewed [2–4].

The increase in the power of computers has enabled calculations to be performed on more complex models. It is now possible to improve models of systems of macromolecules by including in the calculations information on chemical structure, inhibitory potentials, etc. Some relevant papers have appeared on a series of synthetic polymers [5–9], polypeptides [10], polynucleotides [11] and polysaccharides (e.g. amylase [12] and cellulose derivatives [13]). Another development has been the application of MC and MD methods to studies of condensed polymer systems, directly taking into account the influence of molecules of the medium and of intermacromolecular interaction. The information which is emerging from such studies is giving fresh insights into the behaviour of polymer molecules and a new field of polymer physical chemistry is developing.

1 DILUTE SOLUTIONS

1.1 Methodological Questions

In a computer simulation of dilute polymer solutions (macromolecule + low-molecular-mass liquid), all particles under consideration are thought to be in a hypothetical cell, usually of a cubic form with a given edge length L (Figure 5.1). To eliminate the surface effect, periodic boundary conditions are introduced: the main cell is surrounded by similar cells. It is assumed that there is free particle exchange through the cell walls between neighbouring cells. In addition, in dynamic methods, the particle entering the main cell from a neighbouring cell is assumed to have the same momentum as one crossing the wall to leave the cell. It is also possible to introduce periodic conditions in a different way. Molecules are not placed within a cube, but on the surface of a hypersphere with a given radius and dimensions that are one dimension unit greater than those of the system modelled [14].

Among polymer systems, dilute solutions are the most inconvenient systems for computer simulation. Although a polymer molecule is the principle object of interest, most of the steps in the calculation involve the motions of solvent molecules. This is because the number of the latter within a cell is much higher than that of polymer links. Indeed, to exclude the interactions between the chains, it is essential that the length of the cell side should exceed the maximum length of the macromolecule chain. Consequently, when simulating even short chains, e.g. of 20 links with a contour length of $\sim$20 units, the cell on which the calculations are

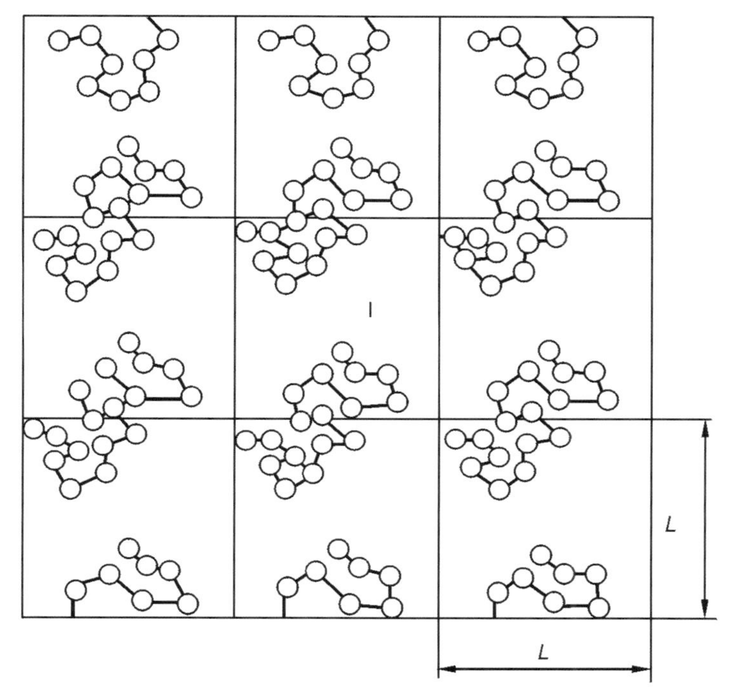

Figure 5.1 Plan view of the hypothetical cell (I) with periodic conditions on the boundaries. The cell contains one chain; solvent particles are not shown

based will contain ~100 particles of solvent if the density of the liquid is about 1 unit. Calculations on such systems are near to the limits of the capacity of the majority of modern computers. In addition, as in a conventional experiment on dilute solutions, the magnitude of the effect of the solute is very small. New methods to improve the calculations will first be discussed briefly. The specific peculiarities inherent in the mathematical simulation of polymer systems will also be considered.

Consider a system of N particles interacting with a given pair potential $f(r_{ij})$, where $r = (r_1, \ldots, r_N)$ are the radii vectors of the particles that determine the configuration of the system. The potential energy $U(r)$ is then given by

$$U(r) = \frac{1}{2}\sum_{i}^{N}\sum_{j}^{N} f(r_{ij})$$

In the classical Monte Carlo method developed by Metropolis *et al* [15, 16], the transitions between configurations are simulated as follows.

A particle i is selected randomly. For a displacement in an arbitrary direction $(r_i \rightarrow r_i')$ for a distance $\leq \beta$, the transition into a new configuration $(r \rightarrow r')$ occurs with a probability, p, given by

$$p = \min\{1, q(r, r')\} \tag{5.1}$$

where $q(r, r') = \exp[(U - U')/kT]$ and U and U' are the respective potential energies. Equation (5.1) means that p is equal to unity or to $q\,(r, r')$, whichever is the smaller. The probability that the system does not change its configuration is given by $(1-p)$. The process of random selection of a particle and estimation of the probability of displacement (step) is performed many more times, 10^3–10^5 steps per particle. The characteristic of interest is averaged over all the steps. The estimation obtained for a very large number of steps approaches the real average value for the canonical ensemble.

A modification of the Metropolis method has been suggested [17, 18] so as to simplify the calculations. The particle selected is assumed to move with higher probability in the direction of the momentary force acting on this particle from rest. In this algorithm, called 'the method of displacement by force,' the new configuration is assumed with probability p given by

$$p = \min\left\{1, q(r, r') \cdot \frac{\boldsymbol{T}(r', r)}{\boldsymbol{T}(r, r')}\right\} \tag{5.2}$$

where $q(r, r')$ has the same significance as in Eq. (5.1), and the value of $\boldsymbol{T}(r, r')$, relating to the transition $r \rightarrow r'$, is given by

$$\boldsymbol{T}(r, r') = \begin{cases} C\exp(\lambda \Delta r F_i/kT) & \text{at } \Delta r \leq \delta \\ 0 & \text{at } \Delta r > \delta \end{cases} \tag{5.3}$$

The corresponding expression for the reverse transition, $\boldsymbol{T}(r', r)$, has a similar form. The terms in these expressions have the following significance:

$$\Delta r = (r_i' - r_i)$$

This is the displacement of the ith particle within the limits, δ, of a particular region.

$$F_i = -\partial U(r)/\partial r_i$$

This is the force acting on the ith particle in configuration r. λ is a parameter and C is the setting constant, another parameter which is cancelled out in Eq. (5.2). Parameters are selected so as to maximize the mean-square displacement of the particles.

Equations (5.2) and (5.3) are consistent with the fact that particles are mostly displaced in the direction of the force *F*. If λ is made equal to zero then this scheme becomes identical with that put forward by Metropolis *et al.* [15].

Some authors [19–21] favour an MC algorithm that is formally similar to the method of Brownian dynamics and is called SMART [19]. If an *N*-link chain is put into a viscous medium then the motion of any *i*th particle (of the link) can be described by a Langeven equation [22]:

$$m_i \frac{d^2 r_i}{dt^2} = -m_i \gamma_i \frac{dr_i}{dt} + F_i + R_i \qquad i = 1, 2, \ldots, N \tag{5.4}$$

where m_i and γ_i are the mass and friction coefficients, respectively, F_i is the force acting on the *i*th particle due to the rest of the particles of the chain and R_i is a casual force reflecting the chaotic jerks from the molecules of medium. In real situations, Eq. (5.4) should include the hydrodynamic forces, but in highly viscous media ($p \approx 1$) they are absent. In this case the inertia term in the left-hand side of Eq. (5.4) can also be omitted. The following algorithm of motion is then obtained:

$$r_i' = r_i + \Delta t D_i F_i / kT + X_i \qquad i = 1, 2, \ldots, N \tag{5.5}$$

where $r_i' - r_i$ is the distance covered by the *i*th particle during time $\Delta t, D_i$ is its diffusion coefficient, given by

$$D_i = kT / m\gamma_i$$

and X_i is the casual displacement corresponding to force R_i and possessing a normalized Gaussian distribution

$$W(X_{i\alpha}) = (4\pi a_i)^{-1/2} \exp(-X_{i\alpha}^2 / 4a_i) \qquad \alpha = x, y, z \tag{5.6}$$

where a_i is a given parameter and equal to $\Delta t D_i$.

In the Brownian dynamics approach, the equation of motion (5.5) is taken to describe a succession of definite transitions between states. In MC method, the new position r' in Eq. (5.5) is accepted or rejected with a probability, $\boldsymbol{T}(r, r')$, that is determined by Eq. (5.2), where

$$\boldsymbol{T}(r, r') = (4\pi a_i)^{-3/2} \exp\{-[(r_i' - r_i) - a_i F_i / kT]^2 / 4a_i\} \tag{5.7}$$

There is a corresponding expression for $\boldsymbol{T}(r', r)$. Compared with the MC scheme developed by Metropolis *et al.* [15], algorithms of the 'displacement by force' and SMART type result in a 2–3-fold increase in the probabilities of transition from one configuration to another when compared with the probability given by Eq. (5.1).

Lastly, there is the new approach to MC simulation of the chains that is based on the technique of functional integration. Contrary to the usual consideration of a system involving a chain as an entity of discrete force centres (links), it was suggested [23] that a macromolecule should be described as a continuous line which is defined within the functional space by a Fourier series. At an arbitrary selection of the coefficients of the series the line changes its form (i.e. experiences 'conformational rearrangements'), and the transitions between the states are performed according to the usual MC scheme.

When polymers are studied by the MD method, there is an important question of how to describe the valency bonds. There are two alternatives: (1) to simulate bonds by any smooth function (e.g. square) or (2) to consider bonds as absolutely rigid limitations, e.g. to simulate by a δ-function.

In the first case, the standard procedures of mathematical calculations [24] are applicable to integrate the equations of motion of the type

$$m_i d^2 r_i / \mathrm{d}t^2 = F_i$$

These schemes enable one to find the coordinates and impulses on the particles after a finite time periods Δt. However, owing to high 'rigidity' of the valence potentials, parts of the chain are characterized by high-frequency vibrations that can only be taken into account when there are very short integration steps, Δt. This reduces the length of the trajectories that are accessible for calculation. That is why exclusion of the respective degrees of freedom from the calculations when the chemical bonds are considered to have geometric limitations has an important effect on the calculations.

Special procedures for correcting the bond lengths at each step are used within the framework of such a simulation method. One of the correction procedures was suggested by Balabaev [25]. The other widely used procedure, developed by Ryckaert *et al.*, has been named SHAKE [26]. A further development of the SHAKE algorithm has been named RATTLE [27]. It is the existence of such correction procedures that constitutes the specificity of the mathematical simulation of polymers by the MD method. They are included without any corrections or changes in the BD method that is based on numerical integration of the Langeven equations. Some papers [28–30] contain new algorithms which have been applied in calculations by the BD method. These include calculations in which have been taken into account hydrodynamic interactions [29] and the effects of casual force lag [30]. Both dynamic methods yield the phase trajectory from which the time-averaged equilibrium characteristics and time-dependent characteristics can be determined.

There are two general problems to bear in mind when considering aspects of computer simulation. One concerns the interrelation between different computer methods and the other the relationship between computer simulations and direct physical observations. As mentioned previously, the principles constituting MC and

MD simulations are radically different. The first method relies on the probability principles of statistical mechanics and the second on the purely deterministic basis of the classical mechanics of Newton. The BD method combines a stochastic and determinate description and occupies an intermediate position.

The postulate that equilibrium properties that are obtained by statistical mechanical averaging on the ensemble are equivalent to time averages of dynamic systems is the ergodic assumption. If the averages were based on an infinite number of particles, the equivalence could not be questioned. However, computer calculations can only be carried out on a limited number of particles but it can be shown that differences in the two kinds of averages for N particles differ by the order of $1/N$.

There is another uncertainty about the interrelation between the computer methods that arises from the kind of conceptual model used for the dynamic changes postulated in the Monte Carlo process. The scheme put forward by Metropolis *et al.* [15] (or its modifications) reproduces an evolution process in which the system, step by step, goes from one state into another, the neighbouring states being strongly correlated. Of course, such evolution itself implies certain artificial (stochastic) kinematics, and does not reflect the real alteration of the states of the system during real time, in contrast to the molecular dynamics approach. However, it is possible to define conditions under which the MC method yields a correct description of the kinematic behaviour.

Meakin *et al.* [31] have shown that MC calculations in the limit when the size of the step, δ, tends to zero are equivalent to BD calculations when the coefficient of friction is much greater than unity. Under these conditions, the inertia-less motion of particles follows the Langeven equation (5.5) and the Monte Carlo time-scale and the number of steps per particle, s/N, is connected with real time t by the relation $s/N \sim \boldsymbol{T}t/\partial\delta^2$. In practice, the two methods of calculation can often be taken to be equivalent for steps of finite size. The two methods should usually give closely compatible results when applied to the same model. This provides mutual checking of calculations.

The basic problems of computer experiments arise when formulating a model of the object to be studied. First the rules describing the interaction of particles with each other or with external fields must be selected. One of the problems is deciding how to take into account the valence forces at the chemical junction of atoms. Various types of valence and non-valence potentials, which only approximate the real interactions, are used for this purpose. It is the selection of these that determines the degree of the correspondence between physical reality and computer experiments. One possibility is to tackle the problem in reverse. In this case the potentials are first adjusted so that computer predictions are consistent with existing experimental data. These potentials can then be used to gain more detailed information about a system which cannot be found by direct experiment.

On the other hand, since the numerical experiments yield strict results for a given model within the limits of the statistical error, they can be considered as a standard for checking the theories dealing with the analogous model and the approximations on which the analytical considerations are based. Such a way of stating the question is probably the most methodologically substantiated one. However, one should bear in mind for both cases that any modern computer is only able to operate with an immeasurably smaller number of particles than in real systems, where $N \approx 10^{23}$. This is the main limitation. Usually calculations are carried out for $N \approx 10^2 - 10^3$ and $t \ll 10^{-9}$ s.

Small systems, such as molecular clusters or isolated macromolecules, are of interest in themselves. The macroscopic properties of large systems are usually interpreted by computer methods on the basis of the simulation of a small number of particles. This is not a trivial problem. The greatest difficulties are found when systems are studied near the phase transition points. Large-scale fluctuations with low frequencies appear. The analysis of such problems will not be considered in detail here but has been discussed by others [16, 24].

It is essential that, when using periodic boundary conditions for extrapolation, the length, L, of a cell side ($L \approx N^{1/d}$ in d-dimensional space) should considerably exceed the correlation radius of the fluctuations of the medium. If the calculation is not performed near the phase transition point, then for the usual short-range potentials this requirement is usually fulfilled at $N \approx 10^2 - 10^3$, and the error due to the finite size of the cell is of the order $1/N$ or $1/L^d$.

1.2 Equilibrium Properties

High-molecular-mass compounds cannot be transferred into the gas phase. As a consequence, the influence of the solvent on their conformation and motion can only be considered in principle. This is very important from the methodological point of view. It has a bearing on the validity of computer calculations in which the solvent is not considered explicitly, but taken into account either through the mean force potential of the interaction between links, or as a continuous liquid medium giving rise to casual jerks.

The MC method has been used [32] to investigate a system consisting of one *n*-butane molecule and 127 tetrachloromethane molecules with an overall density of 1.61 g cm^{-3}. The interaction between the particles was described by a Lennard–Jones potential. The inhibition of internal rotations around the central C—C bond in butane was described by a three-barrier cosinusoidal potential $V(\Psi)$. The transitions between configurations were performed according to the scheme of Metropolis *et al.* [15]. The analysis of distribution of the internal rotation angles Ψ shows (Figure 5.2) that these distributions are practically identical both for dissolved and for isolated (in rarefied gas) molecules of *n*-butane. The most probable values in both cases are situated near 0 and $\pm 120°$. The fractions of *trans*-isomers also were found to be similar: 0.61 and 0.65.

It is therefore evident that the solvent does not significantly influence the configuration of *n*-butane within the limits of precision of the calculations. The

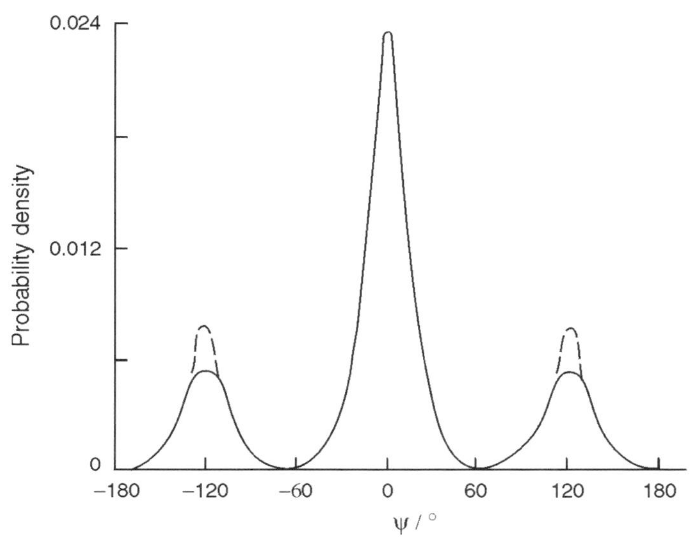

Figure 5.2 Function of the distribution on the values of torsion angle ψ of the central C—C bond for butane at 218 K [32]. Continuous line, gas; dashed line, solution in CCl_4; *trans* conformation corresponds to angle $\psi = 0°$

same conclusion was reached when a dilute solution of *n*-butane was simulated by the MD method [33].

Calculations for *n*-butane do not adequately reflect the properties of long-chain compounds. That is why computer experiments on the simulation of chains with a greater number of links are of special interest. There are various examples in the literature of the use of the MD for this purpose [34–41]. The first papers of this type [34, 39] described investigations on a nine-link chain ($N=9$) within a cell with the number of particles of solvent, N_s, equal to 855. The interaction between any two particles of the system was described by the Lennard–Jones potential:

$$f(r_{ij}) = 4\varepsilon_0[(\sigma/r_{ij})^{12} - (\sigma/r_{ij})^6] \quad (5.8)$$

where r_{ij} is the distance between particles i and j and ε_0 and σ are the Lennard–Jones parameters.

Reduced units are used in the MD method: length is expressed in units of diameter of the molecule, d, rate in units of ε/m and time in units of τ, where $\tau = d(m/\varepsilon)$ and m is the mass. Bruns and Bansal [34, 39] assumed the mass to be the same for all the particles. The chain that they considered had bonds of strictly fixed length, but the angles between the bonds and the angles of internal rotation were not limited. The average numerical density, ρ, of the system was given by

$$\rho = (N + N_s)/L^3 = 0.85$$

The mean square of the distance between the chain ends, $\langle r^2 \rangle$, and mean-square radius of gyration, $\langle s^2 \rangle$, were found by averaging on the trajectory of extending over a time of 142τ. In addition, the calculation of the static structural factor $s(Q)$ at $|Q| \ll 40/\sigma$ was carried out. For comparison, all the calculations were repeated for the same system but solvent was not taken into account. The average sizes of the chains appeared to be similar in each case. In addition, functions $s(Q)$ were practically identical. Small differences were observed in the region of small values of Q when $Q \ll 1.5/d$. These corresponded to the links that are situated far from each other. In addition, the characteristics of the average form of the polymer tangle (i.e. the ratios between the orthogonal components of theradius of gyration) were not noticeably different.

Rapaport obtained a similar result [36] when he used the MD method to investigate freely jointed chains having 5, 10 and 20 links in the presence of a solvent at $\rho = 0.2$ or 0.3. It was shown that the values of $\langle r^2 \rangle$ and $\langle s^2 \rangle$ agreed well with the data from his previous calculations [42] in which the solvent was not taken into account.

Jeon *et al.* [41] carried out calculations on an MD simulation of a dense system (ρ=0.85) of 246 particles of solvent and one 10-link chain with fixed skeleton angles and bonds and with inhibited rotation. The calculations took into account strong attraction between all the particles of the system. It was shown that values of $\langle r^2 \rangle$ and $\langle s^2 \rangle$ increased somewhat compared with results of calculations using a purely repulsive potential. Perhaps this effect is due to the decrease in the total pressure within the system.

The calculations described refer to comparatively short chains. At first sight it is reasonable to doubt whether they are very relevant to the properties of polymers with a much larger number of links. However, application of the MC method in combination with methods based on renormalization group calculations [43] showed [44] that calculations based on short chains can give information about long chains. As the length of flexible chains with strong interactions is increased, properties tend to asymptotic values at relatively short chain lengths. Even if the number of links is between 8 and 16, freely joint non-crossing chains satisfy the relationship

$$\langle r^2 \rangle \sim N^{2v} \tag{5.9}$$

where $v = 0.589 \pm 0.003$. This value practically coincides with the limiting value. The most reliable calculation of v that has been made [45] is 0.588 ± 0.001 at $N \to \infty$. A value for v of 0.586 ± 0.004 obtained for polystyrene ($M_w \leq 5.6 \times 10^7$) [46] is the most reliable value obtained by direct experiment.

An inequality $N \gg (b/\sigma)$, is the criterion of asymptotic behaviour, where b is the bond length and σ is the effective rigid diameter of link. For the chains considered in the numerical experiments [34–42] this condition is fulfilled, since in all the cases $b > 1$. That is why even relatively short chains should reflect correctly the equilibrium properties of polymers.

As discussed above, computer simulation methods show that, in dilute solutions, a low-molecular-mass neutral solvent at the usual densities and pressures has little influence on the conformation of a flexible polymer chain. The interactions between the links themselves play more noticeable a role. That is why the use of computer-based models for simulation of the equilibrium properties of a polymer is justified even if the solvent is not considered in explicit form but is taken into account in the selection of the potentials of the interaction between links.

If the effective potentials are such that the interaction between the links is characterized by predominantly repulsive forces, and the macromolecule swells and increases its volume relative to a θ size of r ($r \approx bN$), then it follows that a thermodynamically good solvent has been simulated. On the other hand, if the tangle is compressed relative to r_θ due to a strong attraction between the links, and if it changes into a globular state, then it follows that a poor solvent has been simulated. This means that the quality of the solvent is worse than that of a θ solvent. It follows that the effective potentials behave as the mean force potentials of the interaction, i.e. they determine the change not in the potential but in the free interaction energy.

Additional problems arise in the computer simulation of polyelectrolytes when there are long-range electrostatic interactions inside the chain. Charged flexible chains with the number of links $N \leq 321$ were studied by the MC method in a hyperspace with 3–7 dimensions [47]. A non-screened coulombic potential was used to describe the electrostatic forces. It appeared that v in Eq. (5.9) is given by

$$v = 3/n$$

where n is the number of dimensions. This is an unexpected result. It agrees with predictions based on the classical theory of Flory [48] but differs from predictions based on renormalization group calculations [43]. The latter predict a value of v given by

$$v = 2/(n - 2)$$

when $4 < n \leq 6$ (n is the is the number of dimensions). When $n = 5$, the value of v is 1/3.

Renormalization group calculations also predict that $\langle s^2 \rangle / Nl^2 \propto N^{2v-1}$. The slope of a plot of a log–log plot of $\langle s^2 \rangle / Nl^2$ against N is therefore predicted to be $2v - 1$. It follows that the slope is predicted to be 1/3 when $d = 5$. This is shown in Figure 5.3 where Monte Carlo calculations of $\langle s^2 \rangle / Nl^2$ are compared with the dependence for a five-dimensional space (RG method). It is possible that in the case of the long-range potential even at $N \approx 100$ the length of the chain is insufficient for a complete realization of the volumetric effects.

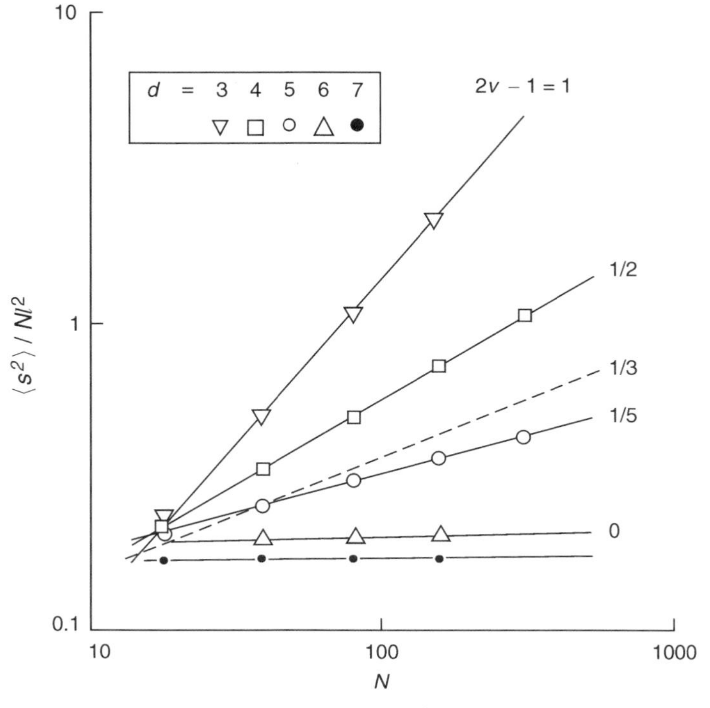

Figure 5.3 Dependence of the normalized mean-square radius of gyration of the charged chain on the number of rigid links N of length l for various dimensions d [47]. The continuous lines correspond to calculations based on Flory's theories. The dashed line corresponds to predictions by the RG method for $d=5$. Data points correspond to MC calculations

1.3 Dynamic Properties

The diffusion coefficients and various time autocorrelation functions (AF) of the form

$$C(A, t) = \langle A(0)A(t) \rangle$$

that characterize, under stationary conditions, the correlation between the magnitudes of the quantity A under consideration at the moments of time separated by interval t, are usually calculated when studying dynamic properties.

The translatory diffusion coefficient D is obtained by integration of AF velocity $C(v,t)$ according to the Green–Coubot equation [22]:

$$D = \frac{1}{2}\int_0^{\infty} \langle v(0)v(t) \rangle \mathrm{d}t$$

where v is the velocity. D can also be obtained from Einstein's equation:

$$\lim[\langle \Delta \boldsymbol{r}^2(t) \rangle \beta]_{t \to 0} = 6tD + B$$

$\langle \Delta \boldsymbol{r}^2(t) \rangle$ is the mean-square displacement of particles during time t and is given by

$$\langle \Delta \boldsymbol{r}^2(t) \rangle = \langle [\boldsymbol{r}(t) - \boldsymbol{r}(0)]^2 \rangle$$

B is a coefficient determining the magnitude of the diffusion process compared with the Markov's process. Angle brackets denote averaging over the time and over all the particles.

The averaging interval of time should be considerably greater than the relaxation time t of the process considered. This is because the normalized AF has the form

$$\langle A(0)A(t) \rangle / \langle A^2 \rangle$$

and the error in calculating AF for a particular value of t is equal to

$$(2t_0/NT)^{1/2}[1 - \langle A(0)A(t) \rangle / \langle A^2 \rangle]$$

Information on small-scale and high-frequency motions corresponding to times of 10^{-10} s can be found from measurement of dielectric relaxation, polarized luminescence, NMR, EPR, etc. Such changes have also been studied by various authors using the MD method [37, 38, 49].

In order to understand the behaviour of polymer molecules one may first consider the simplest anisotropic 'molecule' of two firmly bonded particles ('dumb-bell') moving within a liquid of non-bonded particles. The average square of the displacement of the mass centre of the dumb-bell is a linear function of t, except for a short initial part, the inertia region, where $\langle \Delta r^2(t) \rangle \approx t^2$. As in a homogeneous system, the boundary between the inertia and diffusion regions is determined by the time t where $t \approx \partial/m, \partial$ being the friction coefficient of the moving particle. The diffusion law also controls rotation of the dumb-bell. The rotary AFs of the bond vector b are given by the expression

$$\langle P_i[\cos\theta(t)] \rangle = \exp[-\boldsymbol{l}(\boldsymbol{l}+1)D_r t]$$

where $\cos\theta(t)$ is the cosine of the rotation angle b during time t, P_i is the ith Legendre polynomial, D_r is the rotary diffusion coefficient and $\boldsymbol{l}$ is a vector. It should be noted that the above is correct for the bond length $b < \sigma$, when the contribution from inertia to AF relaxation is small. Otherwise, one characteristic time is not sufficient for the description of the orientational AFs.

There is a strong mutual dependence between the translatory and rotary motions of an anisotropic particle in a liquid [50]. This can sometimes be shown by spectroscopic measurements.

In a dense medium, the more rapid the translatory motion of a particle, the more rapid is its rotation. The correlation is explained by the existence of local structure in a liquid. The diffusion coefficient D for a dumb-bell shaped molecule increases linearly with increase in temperature at constant density of the system. This means that the local viscosity of the medium is determined mainly by the magnitude of the free volume.

In general, one can conclude from the analysis of MD calculations that it is justified to describe the motion of an anisotropic particle in a solvent as that of a Brownian particle in a continuous viscous medium. The precision of such a description increases with increasing differences in mass, size and anisotropic form between the molecules considered and the molecules of the solvent. It should be borne in mind that the translatory transference of a particle and also changes in its speed of rotation are made up of a large number of small jumps, not of a small number of large jumps. There is strong interaction of a particle with its surroundings as it diffuses and its motion is a collective process. Such an idea was expressed by Frenkel [51]. The MD method can reveal many subtle details of the molecular motion in a dense medium.

MD studies on flexible chains in media in which interaction between molecules is described by the Lennard–Jones potential, Eq. (5.8), shows [49] that over a small period of time, t, where $t \leq 0.2\tau$ the links can be considered to move as in a gaseous phase. Over a greater time interval where $0.3\tau < t < 4\tau$ the displacement, Δr, remains small compared with the characteristic size of the chain and is consistent with the relationship:

$$\langle \Delta \boldsymbol{r}^2(t) \rangle \sim t^{\alpha}$$

where $\alpha < 1$. Darinskii *et al.* [37] have shown that $\alpha \approx 1/2$. The occurrence of such a special intermediate regime of motion that is absent for the free particles is due to the inhibitory action of the skeleton bonds. The presence of branches causes additional inhibition and the expression for the displacement must be written as

$$\langle \Delta \boldsymbol{r}^2(t) \rangle \sim t^{\alpha}/k$$

where k is the functionality of a given monomer [38].

The relationship $\langle \Delta \boldsymbol{r}^2(t) \rangle \sim t^{\alpha}$ was also predicted from the model of Gaussian subchains put forward by Kargin, Slonimsky and Rauze [43]. At large values of t the usual diffusion law is always fulfilled but the total length of the chain influences the monomer displacement, $\langle \Delta \boldsymbol{r}^2(t) \rangle t/N$.

A higher velocity of motion of the end links compared with the internal links (the difference is about 10–20% [52]) is a peculiar feature of the chains with rigid bonds. This effect of 'hot ends' is due to the difference in the number of the degrees of freedom.

Relaxation of the spatial orientations of the bond vectors of the chain that is plunged into a solvent was studied by Darinskii *et al.* [37]. Comparison of the results of MD calculation with the analytical predictions from the viscous–elastic model of Herst and Harris has shown that relaxation of the average cosine of the bond rotation angle $\left[\text{i.e. orientational AF of the form } \langle\cos\theta(t)\rangle\right]$ in a real chain is practically the same as the relaxation predicted for equivalent quasi-elastic elements in the Herst–Harris model. Here, the rigid bond of the chain corresponds to the elastic element of the model, and the average angle between the elements of the viscous–elastic model is close to that between the bonds. In addition, the ratio between the average cosines and average squares of the cosines of the rotation angles of the vector of chain bonds appears close to the same ratio for an individual dumb-bell diffusing in the viscous medium. In other words, the characteristic times of the orientational AFs correspond, with good precision, to the expression

$$t_1/t_2 = \boldsymbol{l}(\boldsymbol{l}+1)/2$$

However, the orientational AFs of the macromolecular bonds do not correspond to a simple exponential dependence with one characteristic time t_1 and require a wide spectrum of relaxation times for their description. The same conclusion is reached when modelling an individual chain by the MD method [53]. The reason for this is the mutual dependence of various motions within a macromolecule and is a reflection of its chain structure. It is not connected with the presence or absence of solvent.

Kroon-Batenburg *et al.* [156] have shown that it is possible to estimate the persistence length and other properties of cellulose by MD simulations of a small fragment, taking into account solvent effects. Water was taken to be the solvent. Calculations indicate an upper limit of the persistence length of 145 ± 10 Å. Some folded conformations seem to be indicated when theoretical and experimental data are compared. The behaviour of the cellulose chain was found to be consistent with predictions based on the Kratsky–Porod worm-like chain.

The general results of MD calculations can be summarized as follows. The regularities of the local motion of polymer links in low-molecular-mass solvents are close to those which would occur during motion in a continuous viscous liquid. A solvent behaves like a continuous homogeneous medium with respect to a dissolved macromolecule even when the sizes of a solvent molecule and a link in a polymer chain are comparable. Consequently, the MD method confirms those *a priori* premises that are used when modelling the dynamics of polymers by the BD method. In the latter method the medium is considered to be a continuum and not taken into account explicitly but only through the effective friction coefficient of a link, γ, in a Langeven equation. Assuming that the chain links move like Brownian particles, it is usual to use values of the γ parameter which can be obtained from preliminary MD calculations. Since in the BD method there is no necessity to solve the equations of motion for a large number of solvent molecules, the observation time can be prolonged considerably. This is the main advantage of the BD method. It was used, for example, when studying the conformational *trans–gauche* (T $\rightleftharpoons$ G) transitions in polymers with internal rotation barriers [53, 54]. It was found that conformational transformations occurred mainly due to weakly correlated single-barrier T $\rightleftharpoons$ G transitions. The rate constant for the transformation is given by an expression of the form $K\exp(-E^*/kT)$, similar to the Arrhenius expression. The activation energy E^* is little different from the height of the potential barrier which separates the rotation isomers T and G. Cooperative transitions of types GTT $\rightleftharpoons$ TTG and TTT $\rightleftharpoons$ GTG take place less often.

The motion of a polymer chain will now be considered. The autocorrelation function for the velocity of the mass centre of the tangle has a complicated form. A region of negative values [39] is predicted after a rapid decrease to zero. For $N=9$ this would take place during a time $t \approx 0.2$. Such a phenomenon is associated with a small number of links in a chain, and explained by assuming that the diffusing molecule experiences a certain 'reflection backwards' when colliding with the shell of the surrounding particles of medium. This is only possible during the time in which the local surroundings (cell) round the polymer molecule are preserved.

Comparison of magnitudes of AF for isolated chains and for chains surrounded by solvent is of interest. The relaxation processes become slower in a viscous medium. Relaxation times and also r^2 and s^2 of nine-link chains suspended in a solvent, exceed the corresponding values for the isolated chains by more than by one order of magnitude [39]. The relaxation vector $\boldsymbol{r}$ connecting the end links relaxes the slowest. This is due to the necessity of reorientation in space of entire chains. It has been shown [55] that distance $|\boldsymbol{r}|$ is an aperiodic function of time which undergoes certain relatively small fluctuations. As for isolated chains, these are characterized by regular pulses, the period of which increases in direct proportion to N. A temperature increase causes AF to decrease more rapidly with time. At $t=10$ units of time, ε/k for a freely jointed chain of 10 links decreases quickly to zero during a time $t \ll \tau$. This time corresponds to approximately one vibration period of a pair of particles [35].

Jeon *et al.* [41] investigated the influence of attraction in dilute solution. It was shown that if the attraction between all the particles of the system were taken into account then the apparent rate of decrease in AF was reduced.

The study of the dependence of the characteristic times of relaxation of a polymer tangle upon the molecular mass is important. A standard approach for an isolated chain without hydrodynamic interaction involves the application of a dynamic interpretation of the MC method. Calculations with lattice models yield results that are, to a large extent, determined by the mechanism chosen for the motion of the links. To some extent this results in various artifacts such as a marked dependence of the maximum relaxation time of the tangle, t_{m}, on N. Models developed without making use of the concept of a lattice are preferable.

A study of non-intersecting ring macromolecules described by Pletneva *et al.* [56] involved the MC method together with dynamic scaling principles [57]. As in the case of comparable equilibrium properties, the asymptotic dependence of t_m on N is reached at a relatively low value of N, in this case $N \leq 40$.

As N is increased it was found that the correlation approaches

$$t_m \sim N^\alpha \tag{5.10}$$

The value of α is close to 2.2. A similar relationship was derived when non-intersecting chains (N=5–63) were modelled by use of the SMART algorithm [20]. The concept of a lattice was not used. It follows from Eqs (5.9) and (5.10) that with good precision $\alpha = 2v + 1$. For the Gaussian chains it has been found that $\alpha = 2v$ [43]. The calculation described by Pletneva *et al.* [56] did not reveal any noticeable difference in the dynamic behaviour of simple cyclic macromolecules and their typological isomers with trefoil junctions. Values of t_m were found to be practically the same for the same value of N.

The role of the hydrodynamic interactions can be followed by MD simulation of solutions or studies of an isolated chain on the basis of the appropriate algorithms of Brownian dynamics. Calculations by the MD method of the autocorrelation functions for chains in which $N \leq 20$ and which are surrounded by solvent have been described by Rapaport [36]. It was shown that if the overall density of a system, ρ, is ≤ 0.3 then values of AF show a simple exponential dependence on t_m and are proportional to $\exp(-t/t_m)$. Times, t_m, were also found to satisfy the correlation (5.10) with $\alpha \approx 2.2$. The value of t_m increases if ρ is increased.

An unexpected result was obtained by Fixman [58] when he simulated hydrodynamic interaction by the BD method. Calculations indicated that the ratio between the chain size, r, and its hydrodynamic radius, R_H, depended both on the friction coefficient of a link and on the number of links. This contrasts with the relationship

$$r/R_H = \text{constant}$$

put forward earlier [43]. The variation of r/R_H with N found by Fixman is probably due to the chains ($N \leq 56$) on which he based his calculations being of insufficient length for the ratio to reach a constant value.

1.4 Moderately Dilute Solutions. Scaling and Computer Simulation

Polymer tangles in dilute solution are mostly separated from one another and very seldom come into contact. Their sizes at a temperature $T > \theta$ are consistent with Eq. (5.9). One can define a range of moderately dilute solutions in which the average density of polymer links, ρ, corresponds to the relationship

$$\rho^* < \rho \ll 1$$

ρ^* is the critical value of the average density of polymer links at which the tangles begin to overlap and is approximately equal to $N/\langle r^2\rangle^{3/2}$. It follows from Eq. (5.9) that $\rho^* \approx N^{-4/5}$, i.e. the value of ρ^* is very small at high values of *N*. That is why, in moderately dilute solutions, the local density of links should experience strong fluctuations. According to modern theoretical concepts [43], the radius of correlation of density fluctuations or correlation length, ξ, is the distance r over which the local density of links, $\rho(r)$, differs from the average density, ρ. The correlation length is the basic parameter of such a system and characterizes the dimensions of the regions of heterogeneity.

The ξ parameter can be used as the basis for an elegant and extremely illustrative phenomenological description, the so called blob model. The idea is to enlarge the scale of consideration and make a transition from the usual formulation of the properties in terms of collective interactions of polymer links to the formulation of the same properties in terms of the interactions between blobs. A blob is a part of a chain—a subchain—of N_ξ links with average size ξ. In other words, the enlargement of scale results in the substitution of the repeating link of size l by the subchain of blobs of size ξ. At scales $r < \xi$ the chains preserve their individual properties. Hence for the parts of polymer molecules made up N_ξ links it follows that

$$\xi \sim N_\xi^v$$

by analogy with Eq. (5.9). On the other hand, at scales $r > \xi$ the correlation disappears.

Each chain can be considered to be divided into n blobs where $n = N/N_\xi$ and, on the whole, to have Gaussian sizes where $r \sim \xi n^{1/2}$. It is possible to express such important characteristics as $\langle r^2\rangle$ and Π (osmotic pressure) in terms of ξ or ρ. In the general case at $T > \theta$ for d-dimensional space:

$$\langle r^2\rangle \sim N_\xi^{(2v-1)/v} \sim N_\rho^{(2v-1)/(1-vd)} \tag{5.11}$$

$$\Pi/kT \sim \xi^{-d} \sim \rho^{vd(vd-1)} \tag{5.12}$$

Here, v is given by Eq. (5.9) (or approximately $v = 3/2$) and it is assumed that ξ is a decreasing function of the average density:

$$\xi \sim \rho^{v/(1-vd)} \tag{5.13}$$

Such considerations are based on a more general assumption that in the region of strong fluctuations the thermodynamic and correlation functions are generalized homogeneous functions of their arguments. Such an assumption is the essence of the so-called hypothesis of similarity or scaling. According to this hypothesis, transformation of the linear scale $r \to \lambda r$ (or, in dynamic scaling, of the frequency scale) with an arbitrary similarity parameter, means that strongly fluctuating physical values $A(r)$ should also change similarly $[A(\lambda r) \to \lambda \Delta_A A(r)]$ with a certain similarity index Δ_A. It is easy to establish the relation between the critical Δ_A indices, if some of them are already known, on the basis of the phenomenological concept of scaling. The power correlations (5.11)–(5.13) given above using only the power index v illustrate this. However, calculation of the indices themselves involves microscopic theory.

One of such approaches that takes the cooperative interactions within the system into account is known as the theory of average (or self-consistent) field. Flory's entire classical theory of polymer solutions [48] is based on this approximation. It assumes the equilibrium distribution of density within the system, i.e. coincidence between local $\rho(r)$ and average density, ρ, at any distances. On the basis of this assumption $\langle r^2 \rangle$ would be independent of ρ, and Eqs (5.12) and (5.13) assume the form $\Pi \sim \rho^2$ and $\xi \sim \rho^{-1/2}$, respectively.

Flory's theory, when first published, appeared to be very useful, but as experimental precision increased, we have been faced with an increasing number of contradictory data. As a result, the early 1970s were marked by the appearance of an alternative approach based on a fundamental thesis about the mathematical isomorphism of two systems. These are polymer solutions in the region of strong fluctuations in physical parameters in an external field H near the critical temperature T_c. It was strictly proved [43] that there was an unambiguous correspondence: $H \leftrightarrow \rho$ and $T/(T - T_c) \leftrightarrow N(\text{or } N\xi)$ between 'magnetic' and 'polymer' variables. The theory of the behaviour of magnetic fields is well developed. It is possible to calculate the critical indices at $|T - T_c| \to 0$ (or, in polymer terms, $1/N \to 0$). These indices can be used in the analogous equations for polymers. In some cases, they can be measured experimentally. They can also be estimated from computer-based models which possess an advantage in that there is exclusion of the influence of outside factors, such as heterogeneity of the chemical structure of the sample and its polydispersity, macromolecular association, etc. Computer calculations can reveal several important properties of a system, i.e. the degree of deviation from Flory's classical theory, the limits of applicability of the scaling relationships (5.11)–(5.13) for various values of N and ρ and verification of the ratios between various critical indices found by independent calculations.

Usually, the computer study on the equilibrium properties of moderately dilute solutions is carried out by the MC method. There are two peculiarities when such systems are simulated.

There is a general problem of simulating chains because of the necessity of taking into account a large number of geometrical conditions which are imposed by the skeletal angles and bonds. Generally, at fixed geometrical parameters, independent motions of a separate link of the chain should not be considered. Links at rest should be considered immobile, as is done in the scheme of Metropolis *et al.* [15] for a system of separate particles, where only one randomly chosen particle is transferred in each step. This type of motion can be modelled for chains with rigid bonds in the three ways. If a chain is thought of as a broken line in a cubic lattice, one can imagine jumps of the elementary parts in the form of 'angles' (a fragment consisting of two bonds) or 'gates' (a fragment made up of three bonds). This is the basis of the model of Verdier and Stockmayer [59]. In the algorithm of the 'sliding snake' [60], at each step the end link is imagined to be detached and then attached to the opposite end of the chain. This algorithm is applicable with equal success both to the lattice and continuum models. Finally, if the skeleton angles are not fixed, any internal link *i* can be transferred into a new state by a turn through a random angle round the imaginary axis that connects neighbouring links $i-1$ and $i+1$ (model of Baumgartner and Binder [61]). In all the cases, the probability of the transition between neighbouring configurations is determined by Eq. (5.1).

The use of smooth potentials, but not δ-functions, for the description of the chain structure allows the artificial mechanisms of motion to be abandoned. It is on the basis of this model that the calculations according to the SMART algorithm [19–21], mentioned above, were carried out. In addition, the MC method can be used to simulate chains with strictly fixed bonds (angles), without any *a priori* concepts about the mechanism of motion, provided the correcting algorithms (SHAKE types) that are applied in molecular dynamics are also applied here. However, it is not yet clear if such procedures have any advantages compared with traditional ones [59–61].

A different approach is possible when moderately dilute solutions are simulated without explicitly taking into account the solvent molecules because the average density of the substance within a system is rather low ($\rho \ll 1$). One can use calculation methods based on generating an ensemble of statistically independent configurations. In this case each configuration of the chains are imagined to be built by consecutive increase in the number of links [2–4].

1.5 Monodisperse Systems; Concentration Dependences

Correlation Length Parameter, ξ

By circumscribing a sphere of small radius r around any link of the chain and finding the number of links $N(r)$ within a volume V, equal to $(4/3)\pi r^3$, and then averaging over all the links of the system and over all configurations which are generated, one can calculate the local density $\rho(r)$. This is given by

$$\rho(r) = \langle N(r) \rangle / V$$

At small values of r the overwhelming contribution to $\rho(r)$ is made by the intra-chain density of links. This determines the value of ξ. This is equal to the value of r which satisfies the equation

$$\rho(r) = \rho$$

The concentration dependence of ξ has been studied using the MC method for continuous chains with free internal rotation [62, 63], for lattice chains [64] and for polyethylene solutions [65]. It has been shown that at small values of ρ, ξ is comparable to $\langle r^2 \rangle^{1/2}$ and independent of ρ. If $\rho > \rho^*$ then ξ decreases with increasing ρ. This corresponds to a decrease in density fluctuations. It has been established [62, 63] that for the chains in a volume ($d=3$) and on a plane ($d=2$) the calculated data are in good agreement with the correlation expressed by Eq. (5.13). Hence, for a system of 40-link chains in the concentration range $0.05 \ll \rho < 0.37$, the index x_d that determines the dependence of ξ on ρ in Eq. (5.13) was found to equal -0.65 ± 0.04 at $d=3$ and -1.17 ± 0.09 at $d=2$. For the chains in a cubic lattice, the magnitude of x_3 obtained by other authors [64] is 0.71.

Khalatur *et al.* [65] have demonstrated that ξ is independent of N at $\rho > \rho^*$. This confirms an earlier theoretical prediction [43]. As $\rho \to 1$ the value of ξ approaches the size of the link. It was also shown [65] that for polyethylene chains of length ~0.6 nm at a density of about 70% of the density of pure liquid polyethylene. These data can be compared with an experiment reported by Daoud *et al.* [66] in which a decrease of ξ from 10 to ~1 nm was observed when investigating neutron scattering by deuterated polystyrene at $\rho = 0.01 \to 0.15$. The value obtained is similar to the length of a Kuhn segment which for polystyrene equals 1.7 nm.

Khalatur *et al.* [67] used the MC method to study a moderately dilute solution which had contact with an impermeable wall, and calculated the distribution of link density $\rho(z)$ along the z axis, perpendicular to the wall. The value of ξ was found from the condition that $\rho(z) = \rho$. The power dependence given in Eq. (5.13) was confirmed and the index, x_3, was found to be -0.74 ± 0.15, a value equal, within the limits of error, to the values given above. Determinations of the concentration dependence of ξ are close to each other even though determined by different methods. In all cases values of x_d agree well with theoretical predictions [43]. For many purposes it is a good enough approximation to assume that

$$v = 3/(d+2)$$

for a specific temperature, T. From Eq. (5.13), it follows that $x_3 = -3/4$ and $x_2 = -3/2$. For comparison the value of x derived on theoretical grounds by Flory is $-1/2$.

The correlation length can also be estimated by calculating the intra-chain contribution to the static structural factor. Such calculations were performed by the MC method [68] for a 201-link chains in a lattice. It was found that at $\rho = 0.344$, the magnitude of ξ, expressed in the units of the step length of the tetrahedral lattice, was approximately equal to 10, and the average number of links in a blob, N_ξ, was estimated to be 16.

It follows that computer experiments have shown that moderately dilute polymer solutions can be satisfactorily described on the basis of the blob model. This model is applicable to comparatively short chains from tens to hundreds of statistical segments.

Average Sizes

The dependence of $\langle r^2 \rangle^{1/2}$ on ρ for continuous chains of 15 and 20 links was investigated by the MC method for the first time by Curro [69]. It was found that the tangles that swelled in a good solvent became more compressed as ρ increased. The same effect has been found by various other workers. Sariban *et al.* [70] also simulated a system with a thermodynamically bad solvent at $T < \theta$. In this study the attraction energy, ε, of $-0.5\,kT$ was ascribed to pairs of links drawn together; this resulted in the compression of the tangles in the dilute solution compared to θ sizes. For such system a reverse tendency was observed with increasing ρ. The compressed tangles unwound and their sizes approached $\langle r^2 \rangle^{1/2}$.

A computer study of the functional relationship between $\langle r^2 \rangle^{1/2}$ and N shows [68] that, if the chains are long enough ($N \approx 200$) in a good solvent, the linear dependence, $\langle r^2 \rangle^{1/2} \approx N$, that corresponds to Gaussian chains and agrees with Eq. (5.11), is fulfilled when $\rho > 0.2$. The same result is obtained [71] for a two-dimensional system.

There are various critical analyses of publications on computer studies of the dependence of chain size on concentration [72–74]. It is useful to express such a dependence using a scaling variable, x, that characterizes the degree of the overlap of the tangles and given by

$$x = N^{dv-1}\rho \sim \rho/\rho^*$$

It has been noted [73, 74] that most computer studies have dealt with systems containing insufficiently long chains for the proper determination of the index in Eq. (5.11). Results have differed from theoretical predictions. It is necessary for $\rho \gg \rho^*$ for Eq. (5.11) to be obeyed. In terms of the variable x, this is equivalent to $x \gg 1$. However, since the concentration range in which ρ changes is always limited, the condition that $x \gg 1$ is achieved only in systems containing very long chains. Computer simulation of these is technically complicated. A different approach has been proposed [75]. It follows from the discussion of scaling that the value of $\langle r^2 \rangle$ should satisfy the relationship

$$\langle r^2 \rangle \sim N^{2v} Y^{(2v-1)/(1-vd)}$$

where $Y = \rho/\tau\rho^*$ and τ, the relative temperature difference from the θ point, is given by

$$\tau = (T - \theta)/\theta$$

It is evident that for a fixed value of Y an improvement (i.e. increase) in the quality of the solvent will result in an increase in the ratio ρ/ρ^*. Consequently, by strengthening the interactions in volume one can achieve the condition that $x \gg 1$ without an increase in N. In a computer experiment, it is sufficient to assign a statistical weight $\omega > 1$ to contacting links in order to achieve this purpose.

Calculations [75] have shown that for comparatively short chains (N =40–61) at $\omega = 0.9$, the power index in Eq. (5.11) equals -0.212 ± 0.050. This is close to the theoretical value of 1/4 for three-dimensional systems [43].

The MC method was used [76, 77] to study the lattice model of a solution of a polyelectrolyte. A separate non-intersecting chain of 16 or 64 links, part of which carried a unit charge, was conceived as being placed in a cell with mobile counter-ions. Each counter-ion also carried a unit charge and was considered as an occupied lattice site. The interaction between charges was described by a non-screened Coulombic potential. Using the method of local jumps [59], the authors studied the dependence of the chain size on the fraction of charged links ρ_e and on the average link density. The latter was varied by changing the volume of the cell used for the calculation, keeping other parameters constant. The calculation has shown that the chain swells with increasing ρ_e. Hence the counter-ions surrounding the links are unable to compensate for the electrostatic repulsion between the links. This effect is the most clearly seen in dilute solutions where inter-ionic forces are low and the total electrostatic energy is positive. Polyions then take up stick-like forms. An increase in the inter-ionic forces was modelled by decreasing the size of the cell. When the Debye radius is smaller than that of a chain length, L, the total electrostatic energy becomes negative and the polyion assumes the form of a tangle. In this regime, the role of the electrostatic interactions is insignificant compared with that of the 'usual' interactions in space.

Concentration Dependence of the Average Size of the Macromolecules of Cellulose Derivatives

Cellulose and its derivatives are typical semi-rigid macromolecules possessing one characteristic property: the length of the statistical Kuhn segment (A) exceeds substantially its transversal size (d); this affects the size of the parameter of local symmetry, P, where $P = A/d$. P is considerably greater than unity.

Semi-rigid chains are also characterized by inability to have sharp bends. Because of this, change in their conformation occurs by correlated turns of the links. It is to be expected that, under certain conditions, the oriented ordering in solutions when polymer concentrations exceed a certain critical value will be accompanied by a transition into an ordered state of the liquid-crystalline type.

It follows that the equilibrium rigidity of chains corresponding to these conditions, expressed in terms of Flory's critical parameter f, should satisfy the condition that $f < 0.63$. According to Flory, the f parameter is defined as the fraction of flexible bonds that are twisted relative to rigid colinear sequences and is given by the expression:

$$f = \frac{(z-2)\exp(-\Delta\mathscr{F}/kT)}{1+(z-2)\exp(-\Delta\mathscr{F}/kT)} \qquad (5.14)$$

where z is the coordination number in the pseudolattice model used by Flory and $\Delta\mathscr{F}$ is the difference between the free energies of flexible and rigid conformations in the macromolecule. For extremely rigid rod-like molecules $f = 0$ and for completely flexible molecules $f = 1$ [48]. Unfortunately, experimental determination of the average size of chains in concentrated systems is very difficult. Predictions from modern analytical theories of the liquid-crystalline state are true only for infinitely long chains and cannot be transferred directly to the systems of semi-rigid chains for which contour lengths, L, are often comparable with values of A. This difficulty has been overcome in recent work by Pletneva *et al.* [136]. She carried out a computer simulation by the MC method of a system of semi-rigid chains of nitrocellulose (NC) for which $f = 0.074$. The dependence of $\langle r^2 \rangle$ of NC on the volume fraction of polymer, v, was considered. The calculations were based on the model of the NC chain proposed by Kirste [137], who had successfully used it to calculate the average size and some other characteristics of isolated NC chains at $v \to 0$. In accordance with this model, an NC macromolecule was simulated as a chain of n vectors of length, l, that were consecutively connected with each other to form a definite angle. Mutual positions of the vectors in space, and consequently conformation of the molecule, were determined by $n-1$ angles, and by $n-2$ azimuthal angles of internal rotation, that were assumed to be evenly distributed between 0 and 2π. For any point of contact between two vectors i and j, $|i - j| > 1$, the following interaction potential was introduced:

$$U(r_{ij}) = \begin{cases} \infty & \text{at} \quad r_{ij} < d \\ \varepsilon/kT & \text{at} \quad d \leq r_{ij} \leq 1.5d \\ 0 & \text{at} \quad r_{ij} > 1.5d \end{cases} \tag{5.15}$$

where r_{ij} is the distance between i and j points and ε is an energetic parameter similar to the χ parameter in the theory of polymer solutions of Flory and Huggins. Intermolecular interaction was described likewise. Equalities $d = 0$ and $\varepsilon = 0$ in Eq. (5.15) corresponded to the conditions of a θ solvent. In this case, the average-square sizes of the chains are determined by expression

$$\langle r_\theta^2/l^2 \rangle = n(1+\alpha)/(1-\alpha) - 2(1-\alpha^2)/(1-\alpha)^2 \tag{5.16}$$

where $\alpha = \cos(\pi - \alpha)$. When $\varepsilon = 0$ and $d = l$, an athermal system was simulated. An increase in ε at $d = l$ was equivalent to deterioration of the thermodynamic quality of the solvent.

A system of $m = 2 + 6$ chains interacting with each other within a cubic cell of volume V was considered. The volume fraction of polymer was given by $v = m(n+1)d^3/V$. Periodic boundary conditions were imposed along the boundaries of the cell, so that a macroscopic translation-periodic system was simulated. The characteristics required were averaged over the ensemble of $\Omega \approx 10^3 - 10^4$ independent configurations of the system, i.e. over Ω ways of the arranging the chains in space. Various configurations of the chains were generated using the Monte Carlo method and an algorithm given elsewhere [138].

According to Flory [139], the value of f can be estimated from the relationship

$$\langle r_\theta^2 \rangle / L = (2 - f)/f \tag{5.17}$$

On the basis of this equation the value of f for a real NC chain [139, 140] is for example 0.074 at $\alpha = 0.996$ and $n = 39$. In this case, $\langle r_\theta^2 \rangle / l^2 = 1020$, $(L/l)^2 = 1500$ and $A/l = 57.7$. If the parameter l is the length of a repeating unit of NC and equal to $D/2$, where D is the length of the identity period in the crystal [141] and equal to 1.02 nm [141], then the calculated value of A is 29 nm. This is the same as the experimental value for NC with a degree of substitution of ~2.7 when it is dissolved in ethyl acetate [142]. Taking into account the high rigidity of such a chain, one can expect a weak influence of long-range intramolecular interactions on its conformation. Computer calculations have shown that, in dilute systems at low values of v, the magnitude of $\langle r_\theta^2 \rangle$ is practically independent of ε and close to $\langle r^2 \rangle$. Figure 5.4 shows the dependence of $\langle r^2 \rangle$ on v obtained for an athermal system ($\varepsilon = 0$), i.e. for chains in a thermodynamically good solvent. Average sizes are close to θ sizes at small values of v and $\langle r^2 \rangle$ is approximately constant with increase in v to a value that considerably exceeds the critical concentration for overlap of chains which is approximately $n(1/r)^3 \approx 10^{-3}$. However at higher values of v when $v \geq 0.1$ the average sizes increase sharply with

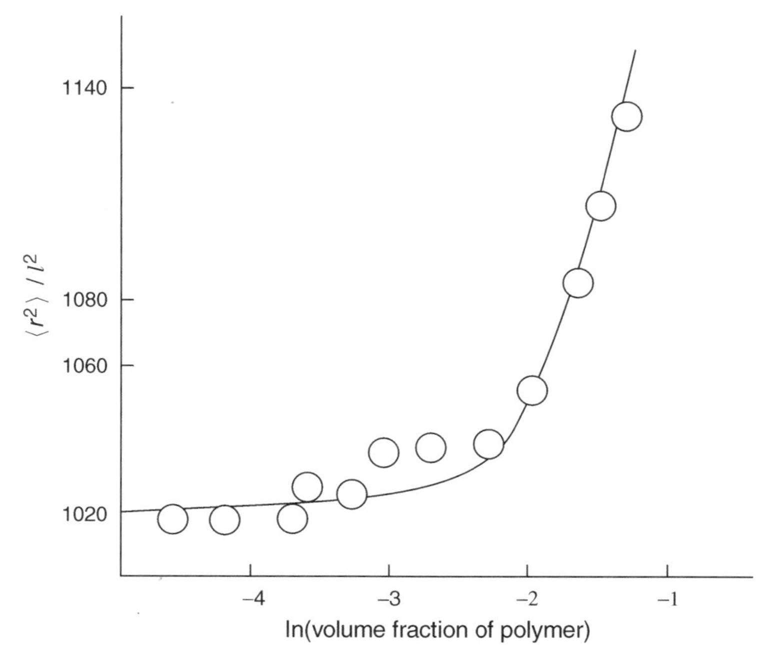

Figure 5.4 Dependence of $\langle r^2 \rangle / l^2$ on the volume fraction of polymer at $\varepsilon = 0$ [143]

increase of v (by ca 12% in the $v = 0.05 \pm 0.26$ region) and are only ~30% smaller than L at $v = 0.3r^2$. It is interesting that the value $v \approx 0.1$ appears equal to the critical value v^*, that in the classical theory of Flory corresponds to the transition of extremely rigid molecules from isotropic into nematic phases. According to the same theory, $v^* = (8/P)(1 - 2/P)$, where P is the degree of asymmetry. In the case of the model considered by the author, when $P = A/l$ then $v^* = 0.13$.

There is thus a clearly defined lengthening of the chains in the model of an NC system in the region of the transition into an ordered state. This effect is an important property of semi-rigid and flexible chains. The latter, in thermodynamically good solvents, are characterized by a decrease in $\langle r^2 \rangle$ rather than an increase with increasing v [143]. In this case $\langle r^2 \rangle \rightarrow \langle r_\theta^2 \rangle$. as $v \rightarrow 1$.

When the influence of thermodynamic quality of solvent on the dependence of $\langle r^2 \rangle$ upon v was investigated [144], it was found that the character of this dependence was close to that in Figure 5.4 when ε was increased in the range $\varepsilon/kT = 0-0.3$. Compared with Figure 5.4, there is a small increase in values of $\langle r^2 \rangle$ (not more than 2%) and small reduction of the region in which $\langle r^2 \rangle = \langle r_\theta^2 \rangle$. Consequently, the effects observed within a system of semi-rigid macromolecules depend weakly on the forces of intermolecular attraction. This is similar to the behaviour of low-molecular-mass liquid-crystalline systems [144] where the main role is played by the repulsive part of the intermolecular potential.

Interaction Between Tangles and Osmotic Pressure

The virial equation for Π, the osmotic pressure, in a dilute solution can be written as

$$\Pi/kT = C/N + A_2C^2 + A_3C^3 + A_4C^4 + \ldots + A_iC^i \tag{5.18}$$

where $C = \pi\sigma^3/6$ and A_i is the ith virial coefficient. The osmotic pressure is also given by the following equation [22]:

$$\frac{\Pi V}{mkT} = 1 - \frac{1}{6kT}\frac{mN^2}{V}\int_0^\infty 4\Pi r^2 g(r)\frac{df(r)}{dr}rdr \tag{5.19}$$

where m is the number of chains in a system, $g(r)$ is the 'molecular part' of the radial function of link distribution and $f(r)$ is the inter-link potential.

The temperature dependence of A_2 has been studied, and the inversion of the sign of A_2 considered elsewhere [78, 79]. The θ point corresponds to $A_2 = 0$. It was found that the critical energy of unitary contact between the links ε at which $A_2(\varepsilon_0) = 0$ depends on the chain length and that $\varepsilon \propto T$. For a given magnitude of N it differs from the value of ε_θ for which there is no swelling of the tangle [80]. The function ε_θ approaches a limiting value of $\varepsilon_\theta^\infty$ since [80, 81]

$$\varepsilon_\theta(N) = \varepsilon_\theta^\infty(1 - N^{e-1})$$

Thus, in the case of finite chains, one should speak not about one θ point, but about a θ region. It should be noted that the notion of a θ region was introduced for the first time on the basis of the analysis of the results of a computer calculation [82]. The strict physical substantiation of this notion was subsequently demonstrated by Khokhlov [83].

Values of Π and A_2 are directly connected with the pair correlation function of the interaction between tangles $G(h)$, that determines the probability of their being at a distance h from each other. It has been shown that, in a good solvent, conformations with a strong overlap of the tangles occur with a probability that differs considerably from zero as $N \rightarrow \infty$ [84–86]. This contrasts with the prediction of the classical Flory–Kribaum theory [48], which predicts zero probability of such states for chains with excluded volume. Confirmation of the results of computer calculations and the shortcomings of Flory–Krigbaum theory have been demonstrated on the basis of simple scaling speculations [84, 85]. More rigorous confirmation has followed by renormalization group calculations [87]. Sharp differences from the Flory–Krigbaum theory are also found when predictions are made about the θ region [4, 88].

The dependence on concentration of the osmotic pressure, Π, has been investigated for an athermal system [89–92]. The interlink attraction was taken into account by Khalatur and Pletneva [93]. It is noteworthy that if the links are simulated as hard spheres with diameter σ, then Eq. (5.15) is transformed into

$$\Pi V/mkT = 1 + 4NCg(\sigma)$$

where $g(\sigma)$ is the magnitude of $g(r)$ at a contact point at $r = \sigma$ and the product $Cg(\sigma)$ determines the average number of contacts per link [84–86].

Calculations for chains without a lattice and with $N \leq 40$ have shown [92, 93] that in a good solvent the correlation holds that $\Pi \approx \rho^X$, where $X = 2.44$. If $d = 3$ it follows from Eq. (5.12) that $X_3 = 9/4$. According to the Flory–Huggins theory [48], $\Pi \approx \rho^2$. The difference is due to the neglect by Flory and Huggins of important correlation effects appearing due to the links being bound together in a chain. Taking the correlations into account it has been shown [93] that the osmotic isothermal compressibility, $\rho^2\beta_T$, is predicted to be given by

$$\rho^2\beta_T \sim \rho^{-0.3}$$

where β_T is the coefficient of relative swelling at constant temperature. If the correlations are not taken into account, then

$$\rho^2\beta_T = \text{constant}$$

An attempt has been made [94] to describe the behaviour of $\Pi(\rho)$ on the basis of integral equations developed in the theory of liquids which are known as the

Perkus–Jevick and the hyper-chain approximations. These approximations were found to give a lower value of the osmotic pressure in the region in which $\rho > \rho^*$. Two-dimensional systems have also been simulated [91, 92]. Calculations of this type indicate that $\Pi \approx \rho^{3.07}$ for an athermal system of continuous chains [92]. This is close to the relationship derived from Eq. (5.2) that $\Pi \approx \rho^3$. In addition, it was also found that $\Pi \sim \varepsilon^{-d}$. Hence, computer simulation of moderately dilute solutions of flexible chains in good solvents largely justifies the use of the scaling method and yields critical indices close to those obtained by use of more recent theories in which systematic deviations from classical concepts [48] are taken into account.

1.6 Polydisperse Systems

Real polymer systems usually possess some degree of polydispersity. The effect of this was analysed by the scaling method [95]. The authors considered the behaviour of a flexible chain of N_1 links that was surrounded by shorter chains of N_2 links of the same chemical composition. It was found that if $N_2 \leq N_1^{1/2}$ then in a good solvent at total density $\rho > \rho^*$, the size of long chains, $r_{1,2}$ in a mixture containing short chains should increase relative to r_1, the size of the same chains at the same density but in the absence of short chains. The relative coefficient of swelling $\beta = \langle r_{2,1}^2 \rangle / \langle r_1^2 \rangle$ depends both on the ratio between N_1 and N_2 and on density ρ.

The MC method was used [96] to investigate a system of continuous non-intersecting chains with free internal rotation. First, the authors obtained the concentration dependence of $\langle r_1^2 \rangle$ for a monodisperse solution of 61-link chains. Then, part of these chains (50%) were substituted by shorter chains, $N_2 = 5$. The density was calculated from the relation

$$\rho = (m_1 N_1 + M_2 N_2)/V$$

where V is the volume of the calculation cell and m_1 and m_2 are the numbers of chains of different kinds within a cell. In a monodisperse system the value of $\langle r_1^2 \rangle$ decreases with increasing ρ. Compression of the tangles was also observed in two-component systems consisting of two kinds of chains. In this case the decrease in $\langle r_{1,2}^2 \rangle$ is evident at higher densities, i.e. $\rho/\rho^* \geq 2$. It was found that at the same density $\langle r_{1,2}^2 \rangle > \langle r_1^2 \rangle$. Consequently the relative coefficient of swelling, β, is $\gtrsim 1$. If some long chains are replaced by short chains then the longer chains become more swollen. This is equivalent to an improvement in the quality of the solvent as far as the longer chains are concerned. However with an increasing value of ρ the absolute sizes of the tangles in a good solvent always decrease but the ratio β/ρ increases. In a system studied by Pletneva *et al.* [96], such an increase continued to $\rho \approx 0.2$. At higher values of ρ it was found that β remained practically unchanged.

The computer simulation of a polydisperse system has been carried out by Bishop *et al.* using the SMART algorithm [97]. A mixture of flexible chains with N_1

$= 70$ and $N_2 = 5$ was considered. The cell contained 10 five-link chains and one 70-link chain with $\rho = 0.3$. It was found that the value of $\langle r^2 \rangle$ of a 'specimen' 70-link chain in the medium of short chains increases by a factor of 1.1 relative to the value in the corresponding homogeneous system. This is close to the results obtained elsewhere [90]. Bishop *et al.* [97] did not study the concentration dependence of β.

It is noteworthy that data from computer calculations [96, 97] agree qualitatively with data from neutron scattering experiments on mixtures of polydimethylsiloxanes [98].

1.7 Polymer Liquids. Conformational Structure of Macromolecules

Two questions are often considered when macroscopically isotropic (liquid, amorphous) polymers are studied by computer methods: (1) what is the difference between the local structure of a liquid consisting of macromolecules with numerous chemical bonds and the structure of a simple low-molecular-mass liquid?; and (2) how does the conformation of chains in a blob change when a polymer is dissolved to form a dilute solution?

Most studies of these problems have made use of the MC method to calculate the sizes of macromolecules that are represented by simple generalized models on lattices or in continuous space. The average size of flexible non-intersecting chains in aggregates at density $\rho \gtrsim 0.6$ are close to θ sizes whatever model is chosen. The dependence of $\langle r^2 \rangle$ on molecular mass is the same as for bodiless Gaussian chains, i.e. in Eq. (5.9) $v = 1/2$. These conclusions confirm the so-called 'Flory theorem' [48]. Such a behaviour is a consequence of the fact that at $\rho \approx 1$ the local density of links does not in practice differ from the average density within the system. That is why density fluctuations are small, and calculations based on the approximation of the average molecular field are justifiable.

Since the late 1970s, numerical computer-based methods have been widely applied to study more complicated systems such as liquid n-alkanes and polyethylene. Direct comparison between results of calculations and experimental data is possible for such systems. The combination of computer methods, physical experiments and analytical theory had led to a better understanding of the systems.

The MC method has been used [99–104] to simulate hydrocarbon chains in the melt. Chemical structure, inhibition and interdependence of the internal rotations were taken into account for every molecule of the system. The interaction of chemically non-bonded groups was described by the Lennard–Jones potential, Eq. (5.8), the parameters of which were selected by experiment. Chains with 8–30 carbon atoms were investigated by Avitabile and co-workers [99, 100] and longer chains up to C_{101} by Khalatur and co-workers [101–104].

Calculations have shown that at a density $\rho \gtrsim 0.7$ (relative to that of pure liquid) the chain sizes assume a constant magnitude and depend only on temperature and chain length, not on ρ. The average-square radius of gyration $\langle s^2 \rangle$ found for the melt

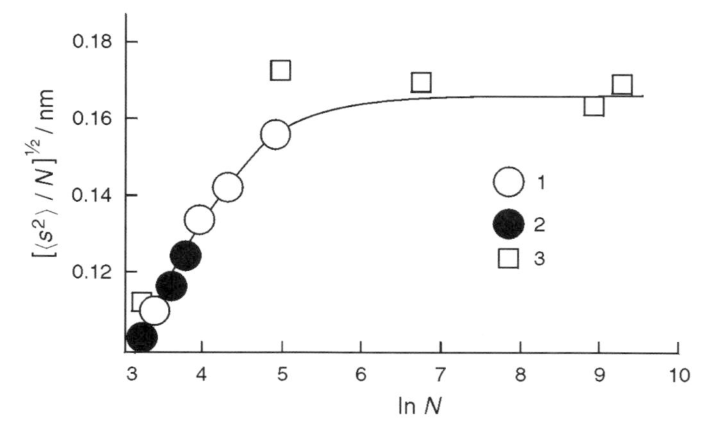

Figure 5.5 Dependence on the number of carbon atoms, N, of $\langle s^2 \rangle$ for polyethylene chains in melt. 1 and 2, results of MC calculations from Refs. 101–104 and 100, respectively; 3, experimental values [105–107]. Continuous line, analytical calculation

are compared in Figure 5.5 with the respective values that are measured experimentally by small-angle neutron scattering [105–107]. Satisfactory agreement between the results is observed in all cases. Values of $\langle s^2 \rangle$ for chains in aggregates are close to $\langle s_\theta^2 \rangle$ values that were calculated analytically by producing matrices for the isolated chains without interactions in space. Consequently, the hydrocarbon chains in the melt are rolled into statistical tangles. It is also important to note that both series of independent calculations [99, 100] and [101–104] were carried out using different algorithms and different computers, but yielded practically identical results.

The mechanism of transitions between *trans* (T) and *gauche* (G) rotary isomers is of great interest when polymers are simulated on computers. Realistic models are needed. Algorithms developed for study of polyethylene are the most convenient. MD calculations for liquid *n*-butane and *n*-decane [108–110] have resulted in the following predictions.

For liquid *n*-butane near the boiling point (292 K), it is found that there is a slightly greater fraction of *gauche* isomers (ρ_G) due to the presence of C—C bonds compared with the gas phase: in the liquid $\rho_G = 0.46$ and in the gas $\rho_G = 0.34$. The average time of T $\rightleftharpoons$ G transition is $\sim$10 ps. Simulation of liquid *n*-decane (27 molecules at 481 K) during 19 ps shows (Figure 5.6) [110] that the average interval Δt between the conformational transitions is $\sim$ 2.1 ps. The fraction of *gauche* isomers for various C—C bonds and times Δt are given in Table 5.1. The relaxation time for the chain as a whole is 2–4 ps. During this time $\sim$15 T $\rightleftharpoons$ G transitions occur. The local conformation of *n*-decane is shown [110] to be practically completely determined by the intramolecular interactions.

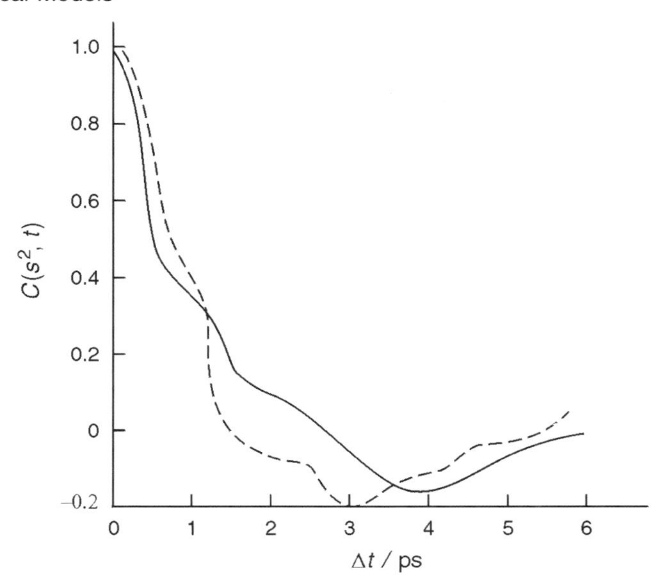

Figure 5.6 Autocorrelation function of the mean-square radius of gyration of decane chain [110]. Continuous line, MD calculation for the melt; dashed line, BD calculation for an isolated chain

Table 5.1 Fraction of *gauche* isomers for various C—C bonds and Δt for the conformational *trans*–gauche transitions in liquid *n*-decane at 481 K [110]

Ordinal number of C–C bond in chain	ρ_G	Δt/ps
2 and 8	0.45	1.9
3 and 7	0.36	2.1
4 and 6	0.34	2.4
5	0.44	1.7

More detailed calculations for repeated identical units have been carried out for liquid oligomers of polyethylene by the MC method [102, 103]. Not only were the amounts of separate T and G isomers calculated, but also those of their different combinations, TT, GG, TG, TTT, etc. Such information has also recently become available from measurements of IR and Raman spectra. Comparison of the results of computer simulation [102, 103] for liquid hydrocarbons C_{21}, C_{51} and C_{101} with the data obtained by IR spectroscopy and with the data from analytical calculations for undisturbed chains has shown good quantitative agreement in all cases. It has also been shown that the local conformational structure of flexible chains in aggregates is little different from the structure of a Gaussian tangle.

These results furnish the most convincing proof of the 'Flory theorem' that have been obtained by computer calculations.

2 CORRELATION IN LIQUID POLYMERS

The correlation radius of the density fluctuations, ξ, is less than 1 nm in liquid hydrocarbons, according to MC calculations [65]. Calculations of the radial distribution functions [99] also show that $g(r) \approx 1$ at $r \geq 0.7$ nm. The first maximum of $g(r)$ in the hydrocarbon liquid, corresponding to the first coordination sphere, is found at 0.4 nm, which is in good agreement with X-ray diffraction measurements. The correlation function of the orientations of intermolecular pairs of C—C bonds has also been calculated [99]. It was found that the orientation correlation disappears completely at distances greater than 0.5 nm. A similar result was obtained [111] by the MD method for polyethylene at density 774 kg m^{-3}. A slight tendency towards perpendicular packing of the segments was observed at $r \leq 0.5$ nm.

It has been established that at $\rho > 0.5$ and in aggregates the polymer tangles overlapped almost completely [86, 112].

The extent of interpenetration of the tangles on a plane is still uncertain. Results of computer calculations diverge. The data from Ref. 114 suggests that there is a tendency for dimeric tangles to segregate. This is supported by other work [43]. However, this conclusion is in contradiction with the other computer calculations [67, 114] and with theoretical consideration [84, 85] based on the same number of molecules.

The relaxation behaviour of hydrocarbons in the melt was investigated by the MD method [108–110]. Calculations [110] of the autocorrelation functions of the tangle size for liquid n-decane have shown (Figure 5.6) that AFs of s^2 and r^2 values decrease to zero during a time $\sim 1.5\tau$, which is 2–3 ps. Consequently, despite the high viscosity of the medium, calculations indicate that the tangle size relaxes quickly. Inclusion of the strong attraction between links into the calculations lowers the rate at which AF decreases [115], but the effect is smaller than it is for a chain in dilute solution. It is interesting that AFs calculated by the MD method [110] were nearly quantitatively reproduced for an isolated chain by the method of Brownian dynamics. The friction coefficient of the methyl was, in this case, selected appropriately [28]. This is shown in Figure 5.6. However, simulation of a dense system by one macromolecule by the BD method is useless when studying phenomena relating to the topological effects that occur due to entanglement of chains in a liquid polymer. The diffusion of chains is an example of such phenomena.

Liquid polymers differ from low-molecular-mass liquids in that there are certain correlations determined by strong chemical bonds between the neighbouring links in the chain. The role of these correlations is important when a van't Hoff's space–time correlation function, $G(r, t)$, is calculated. This function has the form of a

Fourier synthesis and determines the differential part of coherent scattering [22]. It is convenient to divide this function into two parts:

$$G(r,t) = G_d(r,t) + G_s(r,t)$$

where $G_d(r,t)$ is the probability density of finding a different particle at time t at distance r from a given particle and $G_s(r,t)$ is that of finding the same particle. The part $G_d(r,t)$ corresponds to the evolution in time of the radial distribution function. It follows that

$$G_d(r,0) = g(r)$$

G_d and G_s have been calculated [116] by the MD method for a system of 20 flexible 16-link chains at $\rho = 0.8$. The interactions between the links were described by the repulsive part of the potential expressed in Eq. (5.8).

Figure 5.7 a shows the distribution of particles in a liquid of low molecular mass according to the van't Hoff equation. Figure 5.7b shows functions $g(r)$ and $G_d(r,t)$ for different time intervals. The high first peak with height ~ 1.5 at $r/\sigma \approx 1$ in the variation of $g(r)$ (Figure 5.7a) corresponds to the nearest monomer units separated by bonds of unit length σ. Further maxima correspond to different coordination shells. When $r/\sigma > 4$, the value of $g(r)$ becomes very nearly constant. This is also characteristic of simple low-molecular-mass liquids [22]. Hence the main characteristic of the equilibrium structure of a liquid polymer is the existence of an intensive 'valence' peak in the $g(r)$ curve. One might expect that numerous bonds in the chains should have a strong influence on the rate of decay of the density of the pair correlations. However, as Figure 5.7b shows, already at $t > 0.2\tau$ the 'valence' peak in the $G_d(r,t)$ curve 1 has almost completely disappeared. The other peaks are also washed out and disappear gradually with time. Nevertheless,

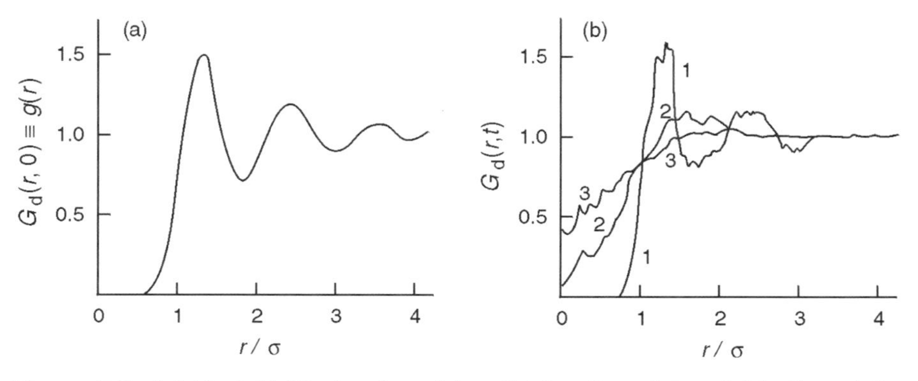

Figure 5.7 (a) Van't Hoff's function; (b) radial function of the distribution of the density of particles in a liquid polymer at different time intervals. Curve 1, $t/\tau = 0.2$; curve 2, $t/\tau = 1$; curve 3, $t/\tau = 2$

even at $t \approx \tau$, the short-range order of the particles of the first coordination sphere is noticeable. At small values of r the region of the forbidden volume is preserved even longer. However, when $t > 4\tau$ the initial structure of $G_d(r, t)$ almost completely disappears.

When $t \gg \tau$, $G_d(r, t) = 1$ and any position of one particle relative to another particle is equally probable. The autocorrelation function $G_d(r, t)$ describing self-diffusion of particles then follows the Gaussian law.

3 REPTATION MOTION

De Gennes [117] proposed a heuristic model called the 'reptation model' to describe the diffusion of flexible chains in dense media where they are strongly entangled. The motion of a chain can be represented using the concept of an effective tube. It is assumed that at any time each chain behaves as if it is contained in a 'tube' of spatial restrictions due to the other chains of the system. A macromolecule diffuses inside this tube due to local conformational rearrangements—'defects' that move predominantly along the length of the chain. The idea of such small-scale, monomeric motion is the basis of the reptation model.

The time, t_R, approximately equal to N units, is the characteristic time during which a chain covers the distance equal to the contour length along the tube. At $t \approx t_R$ the motion of the links obeys the relationship $\langle \Delta r^2 \rangle \approx (t/N)^{\frac{1}{2}}$. For times considerably shorter than t_R ($t \leq t_R/N$), when the average displacement of any link $\langle \Delta r^2 \rangle$ is comparable to the effective diameter of the tube, d_T, a special regime of 'local reptations' should be observed in which $\langle \Delta r^2 \rangle \approx t^{\frac{1}{4}}$ At shorter times, t, the motion of the links is controlled only by the intra-chain bonds, but not by the intermolecular interactions, that corresponds to the usual model of Rauze, i.e. $\langle \Delta r^2 \rangle \approx t^{\frac{1}{2}}$.

Experimental observation of reptations in liquid polymers is extremely difficult. That is why there are no direct unambiguous data up to now confirming or refuting the postulated reptation mechanism. Recently, some workers [68, 97, 118–122] have attempted to find evidence of reptations by the MC method [59, 61]. However, this was not a success when simulating melts. There are probably at least two reasons for the failure. The systems considered were not dense enough ($\rho < 0.7$). For reptation motion to occur, the value of $\langle r^2 \rangle^{\frac{1}{2}}$ should be close to d_τ. If one takes into account that $d_\tau \approx (\sigma\rho)^{\frac{1}{2}}$ (σ is the diameter of monomer), it becomes evident that it is difficult to reproduce the necessary regime in a numerical experiment. Another reason for the failure arises from the fact that when polymers are simulated by the dynamic MC method, the motion of a macromolecule is taken to be due to consecutive rearrangements of small parts of the chains. For chains with interactions in space and rigid bonds, the velocity of the transfer of links along the chain contour is very low. That is why efforts to model the reptation mechanism have not been successful. This is confirmed by the results of Deutsch [120], who could only observe reptation motion by assuming the overlap of monomers inside the same chain. Calculations by Edwards and Evans [121] for one separate macromolecule moving inside a regular net of impediments without interactions in volume and simulated as a rectilinear cubic lattice do not seem to be satisfactory. Apparent reptation motion in the medium of regular impediments seems doubtful.

Various authors [68, 97, 118, 119] have reported that they had managed to simulate the transition from Rauze's regime ($\langle \Delta r^2 \rangle \approx t^{1/2}$) to a reptation regime ($\langle \Delta r^2 \rangle \approx t^{1/4}$) after 'freezing' of the motions of all the chains in melt except for one individual chain. This does not seem to be a very convincing model. The concept of a 'frozen net' would imply that the mobile chain would be surrounded by a closed cell and not by a tube with open ends. This would drastically inhibit its motion.

Reptations of chains in a melt were also studied by MD method [116]. This method is free from faults characteristic of the MC method and thus it yields a more objective description of the dynamic behaviour. Khalatur *et al.* [116] considered a dense system of 20 flexible chains ($N = 16, \rho = 0.8$). They estimated the average squares of the displacement of the links $\langle \Delta r^2 \rangle$. In addition, they found the orthogonal components of the displacement along (tangential component $\Delta r_{\parallel}$) and perpendicular (normal component $\Delta r_{\perp}$) to the chain contour. It follows that $\langle \Delta r^2 \rangle = \langle \Delta r_{\parallel}^2 \rangle + 2\langle \Delta r_{\perp}^2 \rangle$. If $Q(t) = \langle \Delta r_{\parallel}^2 \rangle / \langle \Delta r_{\perp}^2 \rangle$, then $Q(t) = 1$ for isotropic motion. Khalatur *et al.* also considered a separate chain [116] for comparison with a

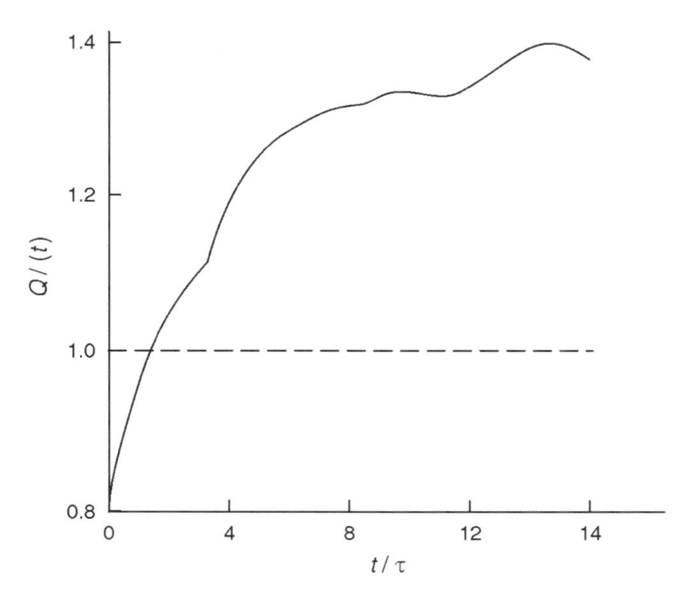

Figure 5.8 Variation with time of the function $Q(t)$ that characterizes the ratio between the displacements of polymer links along and perpendicular to the chain contours. The calculation was made by the MD method for a dense polymer liquid [116]

melt. The calculation has shown that for the separate chain, owing to the presence of rigid bonds, the motion of links in the tangential direction is inhibited. The normal component is preferred. It follows that $Q(t) \leq, 1$. A similar behaviour is also characteristic of chains in a melt over short times, $\Delta t \leq \tau$. In this case the dynamic properties are mainly determined by the chain structure of macromolecules, but not by their interaction. However, at $t > 1.2\tau$, when each link in the melt has enough time to cover a distance $\sim d_T$, the impediments formed by the surrounding chains inhibit motion in the perpendicular direction and cause the links to move predominantly along the chain contour. As a result, $Q(t) > 1$. The variation of $Q(t)$ with time is shown in Figure 5.8. Over much longer periods of time the motion should be isotropic.

Hence the dynamic numerical experiment shows that in liquid polymer there is a predominant type of motion of chain links along the chain contour. This motion is completely determined by the specific inter-chain interactions and can be identified with the local reptations in the melt. The situation is unusual from a general point of view: there is local anisotropy of the motion of particles within a system possessing isotropic structure.

4 PARTIALLY ORDERED SYSTEMS. LIQUID-CRYSTALLINE POLYMERS

In recent years there has been a marked increase in the practical application of polymeric liquid crystals. This has stimulated theoretical investigations of the conditions of occurrence and measurement of physical properties of liquid-crystalline systems. The possibility of the transition into a liquid-crystalline phase depends not on the long length of a polymer chain but on its rigidity, that is characterized by the static Kuhn segment A. If $A \gg \sigma$ (σ is the chain diameter), then orientational ordering can occur under certain conditions. Polybenzamide, poly(benzyl glutamate) and cellulose derivatives are typical examples of mesogenic macromolecules for which $A \gg \sigma$.

There are two basic approaches to the theoretical description of the orientational ordering of macromolecules. One method is that of Onsager [123], which uses virial decomposition of free energy of a rarefied gas of rigid rods. Flory's method [124] is based on the lattice model of a polymer. The modern development of Onsager's theory has been described by Khokhlov and Semenov [125].

Unfortunately, the analytical approach seldom yields numerical results. This is especially true for the most interesting semi-flexible chains, the length L of which is often comparable to A. Hence there is a need for computer simulation of such systems.

4.1 *Equilibrium Properties*

The conditions for the change into a liquid-crystalline state and the conformational changes which occur on the orientational ordering of semi-rigid chains were

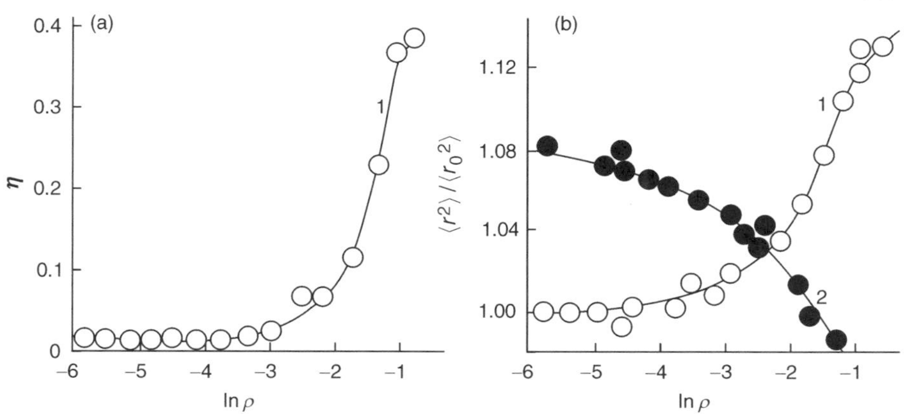

Figure 5.9 (a) Dependence of the orientational order parameter, η, on the density; (b) dependence of the average size of chains on the density (the value of $\langle r^2 \rangle$ is given relative to the value for an isolated chain) [126]. 1, System of semi-rigid chains; 2, system of flexible chains

deduced for the first time by Pletneva *et al.* by the MC method [126]. Chains with free internal rotation and consisting of rigid links of unitary diameter σ and bonded by bonds of unitary length at an angle less than α were considered. The angle α was chosen so that the parameter of local asymmetry, A/σ, was 57.7 and the critical parameter of Flory, f, was 0.074, the magnitude for the nitrocellulose chain [13]. The results of the calculations for 40-link chains are illustrated in Figure 5.9, which shows the dependence on the number-average density ρ of the magnitude of $\langle r^2 \rangle$ and of the orientational order parameter $\boldsymbol{\eta}$. The value of $\boldsymbol{\eta}$ is given by

$$\eta = [3\langle \cos 2\gamma \rangle - 1]/2$$

where γ is the angle between the vector R connecting the chain ends and the director of the sample.

At low values of ρ the parameter $\boldsymbol{\eta} \leq 0.1$, i.e. the system is isotropic. Noticeable changes in $\boldsymbol{\eta}$ and $\langle r^2 \rangle$ only begin at $\rho \gtrsim 0.1$, when these values exhibit a marked increase. The increase in $\boldsymbol{\eta}$ shows that the system changes into the liquid-crystalline state (of nematic type). The transition is accompanied by increasing length and rigidity of the chains, $\langle r^2 \rangle \rightarrow L^2$ as $\rho \rightarrow 1$; such an effect distinguishes in principle between the behaviour of semi-rigid and flexible chains. For comparison, Figure 5.9 shows the concentration dependence of $\langle r^2 \rangle$ calculated for a similar system of 40-link chains possessing high flexibility ($\alpha = 90°$ and $A/\sigma = 1$). In this case, the behaviour is quite different. With increasing ρ, the orientational order parameter does not change and is close to zero, and $\langle r^2 \rangle$ approaches $\langle r_0^2 \rangle$. It was shown [126] that the effects of ordering, that are observed within a system of semi-rigid macromolecules, depend weakly on the attraction forces as in the case of low-molecular-mass liquid-crystalline systems. The main role is played by the repulsive part of the interaction potential.

The results of the computer experiment [126] can be compared with theoretical predictions [125]. For the model considered in Ref. 127, the ratio $L/A = 0.67$. The critical density at the transition point ρ_c, estimated from the maximum of the derivative $d\boldsymbol{\eta}/d\rho$, is 0.23. The corresponding value of $\boldsymbol{\eta}_c$ is 0.37. For the same ratio, L/A, theory [125] predicts $\rho_c = 0.26$ and $\eta_c = 0.41$, where the subscript c refers to critical conditions. Agreement with the computer experiment is very good. It is noteworthy that Flory's theory [124] predicts a lower value for the critical density for the model investigated in this computer simulation [126].

The theoretical approach based on Onsager's method [123, 125] depends on the second virial approximation and its applicability is limited to the region of comparatively low polymer concentrations. By contrast, the lattice model [124] describes the ordering in concentrated systems. MC calculations [127] for a dimeric lattice model have deviated substantially from predictions based on Flory's theory at $\rho = 0.952$. Recently, a similar calculation has been carried out [128] for a trimeric system of chains in a cubic lattice. The chain rigidity was regulated by parameter $\Delta\varepsilon$, which determines the preference of the transform on the lattice (*trans* isomer). Using the 'quasi-reptation' algorithm [60], $\sim 3.5 \times 10^9$ configurations were generated for each $\Delta\varepsilon$. The results of calculations of $\boldsymbol{\eta}$ and of the average fraction of *gauche* bonds, ρ_G, are shown in Figure 5.10. These values can be also obtained within the framework of the lattice theory [124]. This theory predicts that when $N = 20$, in the absence of non-occupied sites on the lattice, the

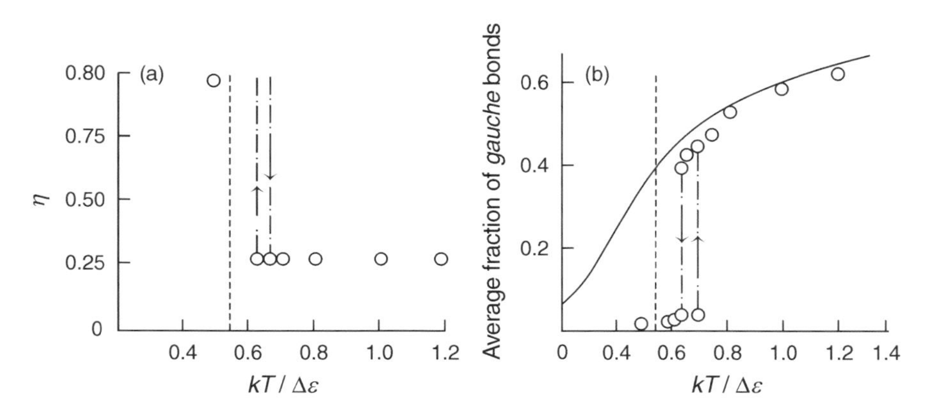

Figure 5.10 (a) Dependence on the reduced temperature of the orientational order parameter, $\boldsymbol{\eta}$ (b) dependence on the reduced temperature of the average fraction of *gauche* bonds corresponding to the lattice model of a liquid crystal. Data points, calculated by the Monte Carlo method; arrows denote the direction of the direct and reverse transitions. Continuous curve, calculated according to Flory's theory; dashed line, reduced critical temperature of transition according to Flory's theory [124]

following relationship holds at the critical temperature of transition:

$$kT/\Delta\varepsilon_c = 0.56$$

Calculation by the MC method shows that there is a sharp transition—the chains become almost completely unwrapped and ordered—as $kt/\Delta\varepsilon_0$ decreases (i.e. while the chain rigidity increases).

As it is seen in the trimeric case, there is a satisfactory correlation between the critical values of $kT/\Delta\varepsilon_c$ that are obtained from theory [124] and from a computer experiment. In addition, a computer experiment reveals the phenomenon of hysteresis: a reverse transition from the ordered into disordered state occurs when $kT/\Delta\varepsilon_c = 0.68$. This cannot be interpreted on the basis of Flory's lattice theory [124].

It is noteworthy that the ordering of rigid chains in the above examples was assumed to take place without any participation of the attraction forces and to be due only to packing effects. In some cases it may be demonstrated that packing effects can be the reason for the orientation ordering of the flexible-chain polymers. For this purpose, it is sufficient to limit the mobility of chains in space. Computer experiments show [129–131] that flexible chains 'bonded' with one of the ends to impermeable surface (e.g. as a result of strong adsorption of the end polar group or of chemical 'grafting') are able to be exclusively orientationally ordered owing to the steric limitations that depend on the degree of filling of the surface. The other possible causes of the transition of flexible chains into the ordered state are either the existence of external orienting fields or the anisotropy of the forces of intermolecular interaction [132].

4.2 Dynamic Properties

There are few papers on the simulation of partially ordered polymer systems by the MC method. Pershin and Skopinov [133] have considered a dimeric system of rigid five-centre molecules and found a tendency for short-lived associates (clusters) to form in which the degree of local order is higher than the average within the system. Association of molecules when approaching each other, in the opinion of these authors, owes its origin to the potential barriers that oppose the shear displacement and stabilize the parallel configuration.

The dynamics of the behaviour of molecules possessing the form of infinitely thin rigid rods was investigated by Frenkel and Maguire [134]. Although the diameter of the rods was assumed to equal zero, they could not intersect with each other. The dependence on the reduced density ρ^* of the AF velocities and the coefficients of self diffusion, D, were calculated. The reduced density was defined as

$$\rho^* = NL^3/V$$

Table 5.2 Coefficient of self-diffusion of rods, longitudinal and diametrical component of its value at various reduced densities [134]

ρ^*	D	$D_{\parallel}$	$D_{\perp}$	ρ^*	D	$D_{\parallel}$	$D_{\perp}$
1	2.40	2.59	2.30	16	0.24	0.50	0.11
2	1.16	1.37	1.05	24	0.21	0.49	0.07
4	0.61	0.82	0.51	32	0.20	0.50	0.05
6	0.44	0.60	0.36	40	0.23	0.64	0.03
8	0.34	0.53	0.25	48	0.34	0.97	0.03

where N is the number of rods of length L within a volume V. The coefficient of self-diffusion was given by

$$D = (D_{\parallel} + 2D_{\perp})/3$$

where $D_{\parallel}$ and $D_{\perp}$ are the longitudinal and diametrical components of the diffusion coefficients, respectively. It was shown that during the time between two collisions the motion is free, not diffusional. At small values of ρ^* the transport properties are well described by the kinetic theory of Enskog [22]. However, at $\rho^* \geq 8$, such a approach is not correct. It is found for this region of densities that the coefficient of the rotary diffusion of the rods decreases in proportion to $(\rho^*)^{-2}$or L^{-2}, in agreement with the theory of Doi and Edwards [135].

An unexpected result was obtained when studying the translational motion. If the system is dense enough, the increase in ρ^* causes an increase in D and $D_{\parallel}$ (Table 5.2), and not a decrease, as should have been expected. Such a behaviour relates to the fact that in a system of infinitely thin rods a change in the velocity component along the rod axis can occur only after a change in the rotation velocity resulting from a collision. However, at high density and long times, the maximum contribution to the respective part of AF velocity $C(v_{\parallel}, t)$ is made only by the rods with a minimum rotary motion. Although the fraction of such particles within a system can be small, nevertheless their role becomes more and more important with increase in ρ^*. That is why the degree of dependence of D and $D_{\parallel}$ of the particles on the maximum component of longitudinal velocity increases with increasing ρ^*.

5 DYNAMICS OF THE CONFORMATIONAL REARRANGEMENTS OF CELLULOSE

Noticeable success has been achieved in recent years in understanding the conformational rearrangements and dynamic properties of complex biomolecular systems: globular proteins, solutions of polypeptides, polynucleotides. The progress in this field is largely due to the development and application of Monte Carlo, molecular and Brownian dynamics methods to study these systems using direct numerical computer simulation. In particular, the dynamic numerical experiments

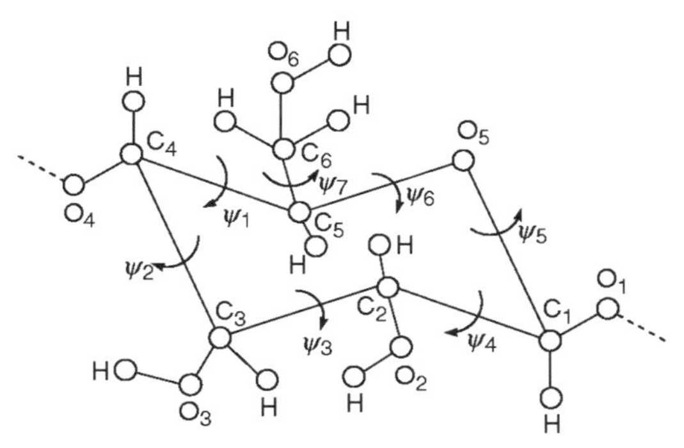

Figure 5.11 A repeating fragment of a cellulose macromolecule

have helped to analyse in detail various types of local motions of atoms in proteins [145, 146].

The first results of the computer simulation of cellulose in order to study the dynamics of small-scale conformational rearrangements were published by Khalatur *et al.* [147]. The calculations were performed for a macromolecule consisting of eight repeating fragments (see Figure 5.11), i.e. of 168 atoms. The chemical bond lengths and valent angles are assumed to be similar to those of cellobiose in its crystalline state [148]. The chemical bonds were considered to be absolutely rigid. The bond angles were held near to equilibrium positions by a harmonic potential:

$$U_\theta(r) = \varepsilon_\theta(r_{ij}^2 - r_\theta^2)^2 \tag{5.20}$$

where r_{ij} is the distance between the end atoms i and j in an i–k–j group of atoms which form an angle θ, r_θ is the equilibrium distance corresponding to a given value of θ and $\epsilon_\theta = 83\,680$ kJ mol^{-1}. Interaction between any pair of atoms i and j that are not bonded to each other and do not form the valence angle was represented by the following potential [149]:

$$(r) = \begin{cases} \varepsilon[(\sigma/r)^{12} - (\sigma/r)^6] & \text{at } r \leq \sigma \\ \varepsilon\left[\dfrac{3(r-\sigma)^2}{(r_c-\sigma)^2} - \dfrac{2(r-\sigma)^3}{(r_c-\sigma)^3} - 1\right] & \text{at } \sigma < r \ll r_c \end{cases} \tag{5.21}$$

The parameters σ and ε characterizing the interaction between H, C and O atoms were taken from Ref. [150]. The radius of the sphere of activity, r_c, in Eq. (5.21) was assumed to equal $3\sigma/2$. To take into account hydrogen bonds O—H$\cdots$O, a function of the '10–12' type [151] was used.

The calculation by the MD method involved the numerical integration of equations of motion, that were written for all N atoms where $N=168$, i.e.

$$m_i \frac{d^2 r_i}{dt^2} F_i + G_i \qquad i = 1, 2, \ldots, N \tag{5.22}$$

where r_i is the radius vector determining the position of atom i, m_i is its mass, F_i is the force corresponding to the potentials of intramolecular interaction and G_i is the force of reactions of rigid chemical bonds.

The BD method allows the investigation of the influence of the viscosity of the medium. In this case, the second term of expression (5.22) included two additional contributions: (1) a random force R_i with a Gaussian distribution of its components, which modelled the chaotic repulsions by the atoms of the medium, and (2) a dissipative contribution $m_i j_i dr_i/dt$, corresponding to effects of friction. Assuming that Stokes' law is fulfilled, the friction coefficient can be written approximately as $\gamma_i = 3\pi\sigma\eta/m_i$ for atom i, where η is the viscosity of the medium. It has been assumed that $\eta = 2 \times 10^{-4}$ N s m^{-2}.

The equations of motion were integrated within Descartes coordinates with a step $\Delta t = 0.002$ ps. The fluctuations of total energy in the MD method did not exceed 0.2%. A modified version of the SHAKE algorithm [152] was used to take rigid bonds into account. The bond lengths were kept constant with a relative precision of 10^{-6}.

After a considerable initial period of relaxation, the trajectories of 10^4 steps were modelled. This corresponded to a time interval of 20 ps. The MD and BD calculations for the parameters selected corresponded to 347 ± 15 and 342 ± 17 K, respectively.

The direct visual analysis of the 'momentary photographs' obtained on the graph plotter shows that in the entire course of the numerical experiment the cellulose macromolecule preserves a strongly stretched conformation. Such behaviour is determined by the geometric peculiarities of the polymer chain and by intramolecular hydrogen bonds. Nevertheless, the atoms experience rather intense thermal motions. To characterize the collective motions of atoms, it is convenient to use dihedral angles of internal rotation ψ.

The BD method shows that the viscosity of the solvent mainly influences these collective types of motions and increases the lifetime of stable conformations. However, the viscosity of the medium does not influence the averaged conformations of the macromolecule. Hence, the chair conformation CI possessing the minimum potential energy and experimentally observable in the crystalline state [148, 153] is the most stable form of the pyranose ring.

However, the conformation CI is not the only possible one. One may follow its inversion to another chair conformation IC during a dynamic numerical experiment. Designations of conformations correspond to those given in Refs. 154 and 155. The transition CI $\rightleftharpoons$ IC is accompanied by the appearance of the intermediate twist forms of the ring. Although such transitions are very rare, they proceed very quickly (during ca 0.1 ps). The 'lifetime' of the intermediate twist forms usually does not exceed a few picoseconds. The calculation shows that angles ψ_5 and ψ_6 rotating around C—O bonds in the ring are the most mobile. As a rule, it is their fluctuations that induce the rearrangements of the entire heterocycle. In this sense, one of these angles is a kind of 'reaction centre.' Considerable mobility of ψ_5 and ψ_6 is explained by the ease of rotation around C—O bonds compared with rotation around C—C bonds. In addition, the oxygen atom, possessing the smallest number of bonds with the other atoms, is the most mobile in the ring. The dynamics of the conformational rearrangements of the ring are characterized by a strong correlation in the change of the torsion angles.

The time of the transitions is calculated to be less than 1 ps when cellulose in vacuum is considered. This time changes a little when the viscosity of medium is taken into account. In both numerical experiments the CH_2OH group preserves a *gauche* conformation during approximately 75% of the calculation time. The preference for this conformation is explained by more advantageous conditions for the occurrence of hydrogen bonds between the hydroxyl hydrogen atom H of the CH_2OH group and the oxygen atoms O-1 and O-3. As a rule, transitions $g^+ \rightleftharpoons g^-$ go through the stage of the formation of a *trans* conformer. However, in rare cases, the transition through the potential barrier in a *cis* conformation ($\psi_7 = \pm 180°$) is possible.

No sharp changes in the corresponding torsion angles are observed when considering the mutual turns of the pyranose cycles around C—O bonds. Owing to the necessity for transference in space of a large number of atoms, the conformational changes of the macromolecule take a relatively long period of time.

REFERENCES

1. Wall F. T., Hiller L. A., Wheeler D. L. *Chem. Phys.* **1954**, *22*, 1036–41.
2. Wall F. T., Windwer S., Gans P. J. *Methods of Computational Physics*, Vol. 1 Academic Press, New York **1963**.
3. Dashevskii V. G. *Organic Chemistry. Results of Science and Technology* VINITI, Moscow **1975**, *1*, 5.
4. Khalatur P. G., Papulov Yu. G. *Computer Experiments in Conformational Analysis of Polymers* Kalinin University, Kalinin **1982**.
5. Lal M., Spencer D. *Mol. Phys.* **1971**, *22*, 649–711.
6. Khalatur P. G., Stepanyan A. E., Papulov Yu. G. *Vysokomol. Soyedin.* **1978**, *20*, 832–36.
7. Khalatur P. G. *Vysokomol. Soyedin.* **1979**, *21*, 2687–91.
8. Dashevskii V. G., Rabinovich A. B. *Vysokomol. Soyedin.* **1983**, *25*, 544–50.
9. Kuchanov S. I., Keshtov M. L., Khalatur P. G. *Makromol. Chem.* **1983**, *184*, 105–111.
10. Karplus M., McCammon J. A. *Annu. Rev. Biochem.* **1983**, *52*, 263–300.
11. Levitt M. *Quant. Biol.* **1983**, *47*, 251–57.
12. Jordan R. C., Brant D. A., Cesaro A. *Biopolymers* **1978**, *17*, 2617–32.
13. Pavlov A. S., Marchenko G. N., Pletneva S. G., Papulov Yu. G., Khrapkovskii G. M., Khalatur P. G. *Vysokomol. Soyedin.* **1984**, *26*, 2319–25.
14. Kratky K. W., Schreiner W. J. *J. Comput. Phys.* **1982,** *47*, 313–18.

15. Metropolis N., Rosenbluth A. W., Rosenbluth M. N., Teller A., Teller E. *J. Chem. Phys.* **1953**, *21*, 1087–92.
16. Binder K., Siperly D., Ansen J. P. in *Monte Carlo Methods in Statistical Physics* (Marchuk G. I., Wikhailov G. A., eds) Mir, Moscow **1982**.
17. Pangali C., Rao M., Berne B. J. *Chem. Phys. Lett.* **1978**, *55*, 413–17.
18. Rao W., Berne B. J. *J. Chem. Phys.* **1979**, *71*, 123–32.
19. Rossky P. J., Doll J. D., Friedman H. L. *J. Chem. Phys.* **1978**, *69*, 4628–33.
20. Ceperley D., Kalos M. H., Lebowitz J. L. *Phys. Rev. Lett.* **1978**, *41*, 313–16.
21. Ceperley D., Kalos M. H., Lebowitz J. L. *Macromolecules*, **1981**, *14*, 1472.
22. Resibois P., De Lerner M. *Classical Kinetic Theory of Liquids and Gases* Mir, Moscow **1980**.
23. Ostrowsky N., Peyraud J.J. *Chem. Phys.* **1982**, *77*, 2081–86.
24. Lagarkov A. N., Sergeyev V. M. *Usp. Fiz. Nauk.* **1987**, *125*, 409–48.
25. Balabayev N.K., *Simulation of Motion of Molecules with Rigid Bonds*. NTsBI AN SSSR, Pushchino **1981**.
26. Ryckaert J. P., Ciccotti J., Berendsen H. J. C. *J. Comput. Phys.* **1977**, *23*, 327–31.
27. Andersen H. C. *J. Comput. Phys*, **1983**, *52*, 24–31.
28. Van Gunsteren W. F., Berendsen H. J. C., Rullmann J. A. C. *Mol. Phys.* **1981**, *44*, 69–95.
29. Allison S. A., McCammon J. A. *Biopolymers* **1984**, *23*, 167–375.
30. Berkowitz M., Morgan J. D., McCammon J. A. *J. Chem. Phys.* **1983**, *78*, 3256–61.
31. Meakin P., Metiu H., Petschek R. G., Scalapino D. J. *J. Chem. Phys.* **1983**, *79*, 1048–54.
32. Bigot B., Jorgensen W. H. *J. Chem. Phys.* **1981**, *75*, 1944–52.
33. Robertus D. W., Berne N. J., Chandler D. J. *J. Chem. Phys.* **1979**, *70*, 3395–3400.
34. Bruns W., Bansal N. in *Short Communications, 26th International Symposium on Macromolecules, Mainz* **1979**, *2*, 981–84.
35. Bishop M., Kalos M. H., Frisch H. L. *J. Chem. Phys.* **1979**, *70*, 1299–1304.
36. Rapaport D. C. *J. Chem. Phys.* **1979**, *71*, 3299–3303.
37. Darinskii A. A., Gotlib Yu. Ya., Neelov I. M., Balabayev N.K. *Vysokomol. Soyedin.* **1980**, *22*, 123–33.
38. Gotlib Yu. Ya., Balabayev N. K., Darinskii A. A., Neelov I. M. *Macromolecules* **1980**, *13*, 602–8.
39. Bruns W., Bansal R. *J. Chem. Phys.* **1981**, *74*, 2064–72.
40. Pavlov A. S., Pletneva S. G., Khalatur P. G. in *Calculation Methods in Physical Chemistry* Kalinin University, Kalinin **1985**, 87–96.
41. Jeon S. H., Oh J. J., Rec T. *J. Phys. Chem.* **1983**, *87*, 2890–94.
42. Rapaport D. C. *J. Phys. A* **1978**, *11*, 213–17.
43. De Gennes P. N. *Scaling Ideas in Physics of Polymers* Mir, Moscow **1982**.
44. Baumgartner A. J. *J. Phys. A* **1980**, *13*, 39–42.
45. Leillou J. C. *Phys. Rev. Lett.* **1977**, *39*, 95–98.
46. Cotton J. P. *J. Phys. Lett.* **1980**, *41*, 231–34.
47. Baumgartner A. J. *J. Phys. Lett.* **1984**, *45*, 515–21.
48. Flory P. J. *Principles of Polymer Chemistry* Cornell University Press, Ithaca, NY **1953**.
49. Grivtsov A. G., Mazo M. A. *Vysokomol. Soyedin. Krat. Soobshch.* **1982**, *24*, 662–65.
50. Ryckaert J. P., Bellemans A., Ciccotti G. *Mol. Phys.* **1981**, *44*, 979–96.
51. Frenkel Ya. I. *Kinetic Theory of Liquids* Nauka, Leningrad **1975**.
52. Balabayev N. K., Grivtsov A. G. *Vysokomol. Soyedin. Krat. Soobshch.* **1981**, *23*, 121–23.
53. Weber T. A., Helfand E. *J. Phys. Chem.* **1983** *87*, 2881–89.
54. Helfand. E., Wasserman Z. R., Weber T.A. *Macromolecules* **1980**, *13*, 526–33.

55. Grivtsov A. G., Mazo M. A. in *Mathematical Methods for Investigations on Polymers* (Livshits I. M., Molchanov A. M., eds) NTsBI AN SSSR, Pushchino **1983**, 32–33.
56. Pletneva S. G., Khizhnyak S. D., Khalatur P. G. in *Properties of Substances and Structure of Molecules* Kalinin University, Kalinin **1984**, 43.
57. Muthukumar M. *J. Stat. Phys.* **1983**, *30*, 457–65.
58. Fixman M. *Physica A* **1983**, *118*, 422–23.
59. Verdier P. H., Stockmayer W. H. *J. Chem. Phys.* **1962**, *36*, 227–35.
60. Wall F. T., Mandel F. *J. Chem. Phys.* **1975**, *62*, 4592–95.
61. Baumgartner A., Binder K. *J. Chem. Phys.*, **1979**, *71*, 2541–45.
62. Khalatur P. G., Pletneva S. G., Papulov Yu. G. *Phys. Lett.* **1982**, *43*, 683–91.
63. Khalatur P. G., Pletneva S. G. *Vysokomol. Soyedin. Krat. Soobshch.* **1982**, *24*, 449–51.
64. Birshtein T. M., Skvortsov A. M., Sariban A. A. *Polymer* **1983**, *24*, 1145–54.
65. Khalatur P. G., Pavlov A. S., Papulov Yu. G. *Russ. J. Phys. Chem.* **1983**, *57*, 1719–22.
66. Daoud M., Cotton J. P., Farnoux B., Jonnink G., Sarma G., Benoit H., Duplessix R., Picot C., De Gennes G. P. *Macromolecules* **1975**, *8*, 804–18.
67. Khalatur P. G., Pletnova S. G., Paplov Yu. G., Pavlov A. S. in *Mathematical Methods for Investigations of Polymers* (Livshits I. M., Molchanov A. M., eds) NTsBI AN SSSR, Pushchino **1982**, 82-87.
68. Kremer K. *Macromolecules* **1983**, *16*, 1632–38.
69. Curro J. G. *J. Chem. Phys.* **1974**, *61*, 1203–7.
70. Sariban A. A., Birshtein T. M., Skvortsov A. M. *Dokl. Akad. Nauk SSSR* **1976**, *229*, 1404–7.
71. Bishop M., Ceperley D., Frisch H. L., Kalos M. H. *J. Chem. Phys.* **1981**, *75*, 5538–42.
72. Khalatur P. G., Pletneva S. G. *Vysokomol. Soyedin.* **1982**, *24*, 472–79.
73. Bishop M., Kalos M. H., Sokal A. D., Frisch H. L. *J. Chem. Phys.*, **1983**, *79*, 3496–99.
74. Okamoto H. *J. Chem. Phys.*, **1983**, *79*, 3976–81.
75. Pletneva S. G., Khalatur P. G. *Russ. J. Phys. Chem.* **1983**, *57*, 2236–40.
76. Brender C., Lax M., Windwer S. J. *J. Chem. Phys.* **1981**, *74*, 2576–81.
77. Brender C., Lax M., Windwer S. J. *J. Chem. Phys.* **1984**, *80*, 886–92.
78. Bellemans A., Janssens M. *Macromolecules* **1974**, *7*, 809–11.
79. Janssens M., Bellemans A. *Macromolecules* **1976**, *9*, 303–06.
80. Baumgartner A. *J. Chem. Phys.* **1980**, *72*, 871–79.
81. Khalatur P. G. in *Properties of Substances and Structure of Molecules* Kalinin University, Kalinin **1980**, 96–102.
82. Rapaport D. C. *Macromolecules* **1974**, *7*, 64–66.
83. Khokhlov A. R. *J. Phys.* **1977**, *38*, 845–49.
84. Khalatur P. G., Khokhlov A. R. *Dokl. Akad. Nauk SSSR.* **1981**, *259*, 1357–59.
85. Grosberg A. Yu., Khalatur P. G., Khokhlov A. R. *Macromol. Chem. Rapid Commun.* **1982**, *3*, 709–13.
86. Khalatur P. G. *Vysokomol. Soyedin. Krat. Soobshch.* **1980**, *22*, 406–8.
87. Witten. T. A., Prentis J. J. *J. Chem. Phys.* **1982**, *77* 4247–53.
88. Pavlov A. S. in *Properties of Substances and Structure of Molecules* Kalinin University, Kalinin **1982**, 83–89.
89. Curro J. G. *J. Chem. Phys.* **1976**, *64*, 2496–2500.
90. Okamoto H., Bellemans A. *J. Phys. Soc. Jpn.* **1979**, *47*, 955.
91. Okamoto H., Itoh K., Araki T. *J. Chem. Phys.* **1983**, *78*, 975–79.
92. Khalatur P. G., Pletneva S. G., Papulov Yu. G. *J. Chem. Phys.* **1984**, *83*, 97–104.
93. Khalatur P. G., Pletneva S. G. *Vysokomol. Soyedin.* **1982**, *24*, 2502–08.
94. Khalatur P. G., Pletneva S. G. *Russ. J. Phys. Chem.* **1982**, *56*, 2316–18.
95. Joanny J. F., Grant P., Turkevich L. A. *J. Phys.* **1981**, *42*, 1045–51.

96. Pletneva S. G., Marchenko G. N., Khalatur P. G. *Vysokomol. Soyedin. Krat. Soobshch.* **1984**, *26*, 287–89.
97. Bishop M., Ceperley D., Frisch H. L., Kalos M. H. *J. Chem. Phys.* **1982**, *76*, 1557–63.
98. Kirste R. G., Lehnen B. R. *Makromol. Chem.* **1976**, *177*, 1137–43.
99. Vacatello M., Avitabile G., Corradinl P., Tuzi A. *J. Chem. Phys.* **1980**, *73*, 548–52.
100. Avitabile G., Tuzi A. *J. Polym. Sci., Polym. Phys. Ed.* **1983**, *21*, 2379–87.
101. Khalatur P. G., Papulov Yu. G. in *25th International Symposium on Macromolecules* **1980**, *3*, 55–56.
102. Khalatur P. G. *Russ. J. Phys. Chem.* **1982**, *56*, 1973–77.
103. Khalatur P. G., Pakhornov P. M., Pavlov A. S. *Vysokomol. Soyedin.* **1983**, *25*, 1667–73.
104. Khalatur P. G., Pavlov A. S., Papulov Yu. G. *Russ. J. Phys. Chem.* **1983**, *57*, 111–14.
105. Dettenmaier M. *J. Chem. Phys.* **1978**, *68*, 2319–22.
106. Lieser G., Fischer E. W., Ibel K. *J. Polym. Sci., Polym. Lett. Ed.* **1975**, *13*, 39–45.
107. Schelten J., Ballard D. G. H., Wignall G. D., Longman G., Schmatz W. *Polymer* **1976**, *17*, 751–57.
108. Ryckaert J. P., Bellemans A. *Chem. Phys. Lett.* **1975**, *30*, 123–25.
109. Weber T. A. *J. Chem. Phys.* **1978**, *69*, 2347–54.
110. Ryckaert J. P., Bellemans A. *Faraday Discuss. Chem. Soc.* **1978**, *66*, 95–106.
111. Weber T. A., Helfand E. *J. Chem. Phys.* **1979**, *71*, 4760–62.
112. Bishop M., Ceperley D., Frisch H. K., Kalos M. H. *J. Chem. Phys.* **1980**, *72*, 3228–35.
113. Baumgartner A. *Polymer* **1982**, *23*, 334–35.
114. Torrie G. M., Barrett J., Whittington S. G. *J. Chem. Soc., Faraday Trans. 2* **1979**, *75*, 369–78.
115. Bishop N., Kalos M. H., Frisch H. L. *J. Chem. Phys.* **1983**, *79*, 3500–3505.
116. Khalatur P. G., Papulov Yu. G., Pavlov A. in *Properties of Substances and Structure of Molecules* Kalinin University, Kalinin **1984**, 3–42.
117. De Gennes P. G. *J. Chem. Phys.* **1971**, *55*, 572–79.
118. Baumgartner A., Binder K. *J. Chem. Phys.* **1981**, *75*, 2994–3005.
119. Baumgartner A., Kremer K., Binder K. *Faraday Symp. Chem. Soc.* **1983**, *18*, 37–47.
120. Deutsch J. M. *Phys. Rev. Lett.* **1982**, *49*, 926–929.
121. Edwards S. F., Evans K. E. *J. Chem. Soc., Faraday Trans. 2.* **1981**, *77*, 1891, 1913, 1929.
122. Needs R. J. *Macromolecules* **1984**, *17*, 437–41.
123. Onsager L. *Ann. N. Y. Acad. Sci.* **1949**, *51*, 627–35.
124. Flory P. J. *Proc. R. Soc. London, Ser. A* **1956**, *234*, 60–73.
125. Khokhlov A. R., Semenov A. N. *Physica A* **1982**, *112*, 605–14.
126. Pletneva S. G., Marchenko G. N., Pavlov A. S., Papulov Yu. G., Khalatur P. G., Khrapkovskii G. M. *Dokl. Akad. Nauk SSSR* **1982**, *264*, 109–15.
127. Baumgartner A., Yoon D. Y. *J. Chem. Phys.* **1983**, *79*, 521–22.
128. Yoon D. Y., Baumgartner A. *Prepr. Inst. Festkörperforsch. Kernforschungsanlage Julich* **1984**.
129. Khalatur P. G. *Vysokomol. Soyedin.* **1982**, *24*, 2061–70.
130. Van der Ploeg P., Berendsen H. J. C. *J. Chem. Phys.* **1982**, *76*, 3271–76.
131. Pavlov A. S., Balabayav N. K., Khalatur P. G. in *Calculation Methods in Physical Chemistry* Kalinin University, Kalinin **1983**, 63–70.
132. Birshtein T. M., Sariban A. A., Skvortsov A. M. *Polymer* **1982**, *23*, 1481–88.
133. Pershin V. K., Skopinov S.A. *Fiz. Tverd. Tela* **1979**, *21*, 946–49.
134. Frenkel D., Maguire J. F. *Mol. Phys.* **1983**, *49*, 503–54.
135. Doi M., Edwards S. F. *J. Chem. Soc., Faraday Trans. 2* **1978**, *74*, 560–70.
136. Pletneva S. G., Marchenko G. N., Pavlov A. S., Papulov Yu.G., Khalatur P. G., Khrapkovskii G. M. *Dokl. Akad. Nauk SSSR* **1982**, *264*, 109–12.
137. Kirste R. J. *Faraday Discuss. Chem. Soc.* **1970**, *49*, 51–59.

138. Rosenbluth M. N., Rosenbluth A. W. *J. Chem. Phys.* **1955**, *23*, 356–59.
139. Flory P. J. *Proc. R. Soc. London, Sec. A,* **1956**, *234*, 60–73.
140. Papkov S. G., Kulichikhin V. G. *Liquid-Crystalline State of Polymers* Khimiya, Moscow **1977**.
141. Bickles M., Segal L.(eds) *Cellulose and its Derivatives* Vols.I and II, Mir, Moscow **1974**.
142. Tsepelevich S. O., Marchenito G. N., Tsvetkov V. N. *Vysokomol. Soyedin. Krat. Soobshch.* **1981**, *23*, 773–76.
143. Flory P. *Statistical Mechanics of Chain Molecules* Mir, Moscow **1971**.
144. De Gennes P. *Physics of Liquid Crystals* Mir, Moscow **1977**.
145. Khalatur P. G., Pletneva S.G., Marchenko G.N. *Usp. Khim.* **1986**, *55*, 679–709.
146. Karplus M., McCammon J. A. *Annu. Rev. Biochem.* **1983**, *52*, 263–300.
147. Khalatur P. G., Marchenko G. N., Pletneva S. G., Khrapkovskii G. M. *Dokl. Akad. Nauk SSSR* **1986**, *291*, 157–62.
148. Chu S. S. U., Jeffrey U. A. *Acta Crystallogr., Sect. B* **1968**, *24*, 830–38.
149. Zimin R. A., Papulov Yu. G., Seregin E. A., Smolyakov V. M., Khalatur P. G. *Calculation Methods in Physical Chemistry* Kalinin University, Kalinin **1985**, 57–87.
150. Scott R. A., Scheraga H. A. *J. Chem. Phys.* **1965**, *42*, 2209–15.
151. Polozov R. V. *The Method of Semi-Empirical Force Field in Conformational Analysis of Biopolymers* Nauka, Moscow **1981**.
152. Ryckaert J. P., Ciccott U., Berendsen H. J. C. *J. Comput. Phys.* **1977**, *23*, 327–41.
153. Pertsin F. I., Nugmanov O. K., Sopin V. P., Marchenko G. N., Kitaigorodskii A. I. *Vysokomol. Soyedin.* **1981**, *23A*, 2147–55.
154. Zhbankov R. G., Kozlov P. V. *Physics of Cellulose* Nauka i Teckhnika, Minsk **1983**.
155. Tarchevsky I. A., Marchenko G. N. *Biosynthesis and Structure of Cellulose.* Nauka, Moscow **1985**.
156. Kroon-Batenburg L. M. J., Kruiskamp P. H., Vliegenthart J. F. G., Kroon J. *J. Phys. Chem. B* **1997**, *101*, 8454–59.

CHAPTER 6

Rheological Behaviour of Lyotropic LC Systems Based on Cellulose and Its Derivatives

The rheological properties of solutions of cellulose and its derivatives are closely related to their phase state. The study of the rheological properties of fibre- and film-forming cellulose derivatives adds to the understanding of the formation of fibres and films and the relationship between physico-mechanical and physico-chemical properties. Rheological properties determine the sort of equipment to be used for processing of cellulose solutions. Methods of stirring, transportation and formation of the products are dependent on these properties.

1 RHEOLOGICAL PROPERTIES OF NON-AQUEOUS SOLUTIONS OF CELLULOSE, ITS ETHERS AND ESTERS

Investigations of the rheological properties of solutions of cellulose and its derivatives indicate a great variety of phenomena both in the state of rest and under the action of shear forces. The latter are important during the formation of films and fibres. Factors which affect the rheological properties include the character of the substituents, their relative proportions in the case of mixed derivatives and the nature of the solvent. Change of temperature may affect the rigidity and permeability of the macromolecules and the structure of the solution. Especially significant is the effect of the degree of polymerization on rheological properties. The viscosity, the solution structure and the properties of films and fibres that are formed from solutions all depend on the degree of polymerization.

There are various publications [1, 2, 6–8, 67, 76, 77] dealing with the rheology of isotropic solutions of cellulose esters of fatty acids [cellulose triacetates (CTA), tributyrates (CTB), acetobutyrates (CAB)] and cellulose nitrates (CN). However, data on rheological properties of non-aqueous solutions of these cellulose derivatives with LC order are extremely limited.

Like solutions of viscose, those of cotton and wood celluloses with different degrees of polymerization are characterized by a classical form of flow curves. An increase in polymer concentration results in increased viscosity. The character of the rheological behaviour of cellulose–polar aprotic solvents–nitrogen tetroxide systems is also typical for systems with isotropic properties.

In general, the dependence of viscosity on polymer concentration in these systems is expressed by the equation

$$\eta = KC^{\alpha} \tag{6.1}$$

where $\alpha = 6$–8. An increase in the shear stress makes the coefficient α decrease to 1. Section I of the curve relating η with C corresponds to dilute solutions. Section II, which is approximately linear, corresponds to concentrated solutions in which the interaction between macromolecules prevails over polymer–solvent interactions.

Various authors have investigated the rheological properties of solutions of cellulose esters and ethers over a wide range of degrees of polymerization and substitution. Measurements were correlated with properties of the corresponding films [1–4]. Work by Klenkova and Khlebosolova [1] on solutions of cellulose triacetate (CTA), tributyrate (CTB) and mixed triacetate–tributyrate (CAB) in various solvents has shown that the character of the flow of solutions and the values of their initial viscosity depend on the chemical nature of the cellulose derivatives and on the length and rigidity of the macromolecules.

The situation varies from solvent to solvent. There is a gradual change in the properties if the acetyl groups of the cellulose macromolecule are successively substituted by butyryl groups. This is related to the change in the skeletal rigidity of cellulose macromolecules. It was noted, for example, that the skeletal rigidity of CTA macromolecules was considerably higher than that for CTB. The skeletal rigidity number for CTA is 5.8 and that for CTB is 1.8–3.4. It was supposed [1] that CAB macromolecules possessed intermediate values of the skeletal rigidity depending on ratio between the acetyl and butyryl groups. It was found that the introduction of other ester groups with larger dimensions into a cellulose macromolecule, in addition to acetyl groups, was a simple way of lowering the initial Newtonian viscosity of solutions. This was especially marked in the case of highly polymeric cellulose derivatives and made it possible to vary the flexibility of the macromolecules and consequently to vary the properties of the films and fibres that are formed from these solutions. The influence of the thermodynamic rigidity of chains on the longitudinal flow of polymer solutions has been considered elsewhere [2, 3]. A comparative study of the rheological behaviour of cellulose diacetate and mixed esters has been reported by Khlebosolova *et al.* [4].

A difference in the characteristics of flow of solutions of polymers with different chain rigidities in DMAA and DMAA + 3% LiCl has been found by Conio *et al.* [5]. Experimental data on longitudinal flow are interpreted from the viewpoint of the superposition of two processes: deformation (unwrapping) of molecular chains and orientation of anisometric molecular formations along the direction of flow.

Klenkova and Khlebosolova [1] and also Kamide and Saito [6] have shown that the viscosity of solutions of cellulose derivatives increases sharply with increase in the average degree of polymerization. As for all polymers, the change in maximum Newtonian viscosity of solutions of cellulose polymers with change of molecular mass obeys the relationship

$$\eta = kM^{3\text{–}4} \tag{6.2}$$

The magnitude of the power index in the above correlation and hence the solution viscosity depend on the shear stress. According to some investigators, the minimum Newtonian viscosity is proportional to molecular mass of polymer:

$$\eta_{\infty} \sim k'M \tag{6.3}$$

It has been shown empirically for dilute and moderately concentrated solutions that the dependence on molar mass of the shear rate (shear strain), in the range of concentration in which there is marked change of viscosity (bend point), obeys the relationship

$$\gamma = aM^{-b} \tag{6.4}$$

where γ is the shear rate and a and b are experimentally determined constants [2].

Polydispersity also influences the curve of flow of a polymer solution. The wider the molecular mass distribution, the less the variation of $\log \gamma$ with change of $\log \tau$ there is, where τ is the shear stress. There also are other ways of expressing the polydispersity, e.g. one of them is based on the plotting of two coordinates according to the equation

$$\frac{\ln(\eta/\eta_{\infty})}{\ln(\eta_0/\eta_{\infty})} = f(\ln\gamma) \tag{6.5}$$

The ordinate in this plot is measured on the scale of probability distribution. This method of estimating polydispersity requires the determination of η_{∞}. This is not always possible. In addition, this method presupposes the symmetry of the flow curve with respect to its bend point, which is not observed in all cases.

There is detailed information in the literature on the rheological behaviour of aqueous solutions of cellulose ethers [7, 8]. The control of the thickening properties of solutions has been discussed, since they are widely used in printing dyes, drilling solutions, anion-exchange membranes and other technological liquid-phase systems. Special attention has been devoted to the influence of additions of dyes, stabilizers and surface-active substances under flow conditions that are close to the regimes of deformation. There are, however, few systematic measurements of the influence of molecular mass, types of substituents and degree of substitution of hydroxyl groups of the cellulose macromolecules.

HOPC solutions exhibit an unusual form of flow curve which is characteristic of systems with liquid-crystalline order and is different from the traditional forms which are peculiar to aqueous solutions of cellulose ethers. This is shown by the way in which η varies with the shear stress τ. A gradual increase in polymer concentration up to a critical value (C^*) results not only in an increase in the absolute values of viscosity, but also in a change in the form of the flow curve [9]. When the HOPC concentration rises above C^*, the viscosity in no longer independent of shear stress over a range of values of the latter. The curve becomes concave, and the system has an inhomogeneous texture.

The flow curves of HOPC solutions in dimethylacetamide [10], water [11] and formic acid [12] are examples. Similar peculiarities, with the appearance of the structural branch of textural heterogeneity of η in the extreme dependence of η on C, are characteristic of lyotropic polymer systems with LC order [13], e.g. of cyanethylcellulose (CEC) in dimethylformamide and in CF_3COOH [14, 15]. The decrease in viscosity with increase in concentration at high concentration of lyotropic polymers has been explained by Kulichikhin [16]. He suggested that it was due to a change in the mechanism of flow (transition to the aggregatic flow) and an increase in the role of orientation phenomena. The stable orientation at shear flow results in the existence of an anisotropic viscosity coefficient, since the internal friction coefficients associated with the motion of rigid macromolecules along and across the major axes differ considerably.

The dependence of viscosity on deformation shear stress is a well known phenomenon for melts and concentrated solutions of polymers. A detailed comparative analysis of the flow curves of low-molecular-mass and polymeric liquid crystals is given in a monograph by Kulichikhin [16]. Figure 6.1a shows

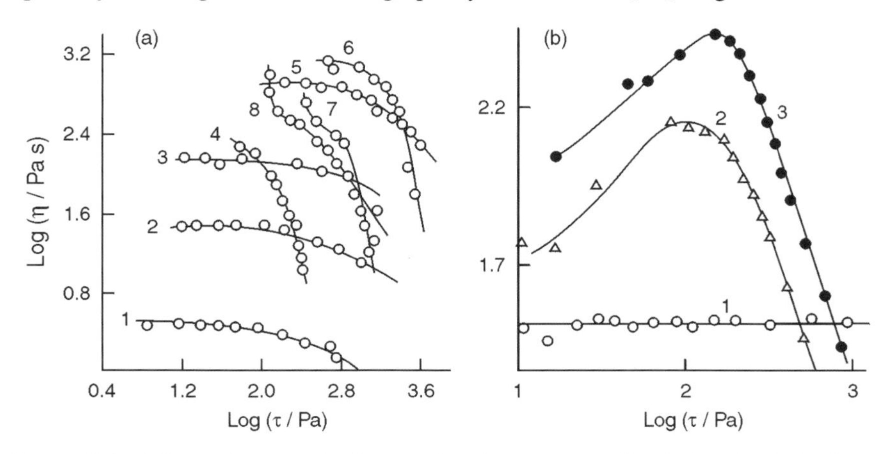

Figure 6.1 Effect of concentration on the flow curves of a lyotropic derivative of cellulose: (a) 1, 10; 2, 20; 3, 25; 4, 30; 5, 35; 6, 40; 7, 45; 8, 50% HOPC solutions in *N*-methylpyrrolidone at 298 K; (b) 1, 17; 2, 21; 3, 24% HOPC solutions in trifluoroacetic acid at 288 K

how the flow curves change with concentration for solutions of HOPC in *N*-methylpyrrolidone. This change is accompanied by an increase in the LC phase. Similar phenomena are encountered in the HOPC–CF_3COOH system (Figure 6.1b)

The author has found marked differences from Newtonian flow in the temperature–concentration regions of phase diagrams which correspond to the existence of two phases: isotropic–liquid crystal and liquid crystal–crystallosolvate (I + LC and LC + CS, respectively).

Various reasons have been put forward for the anomalous viscosity which causes the structural branch of the variation of $\log \eta$ with $\log \tau$. According to Reiner [2], non-linearity of the flow curves is due to the deformation and destruction of particles. He suggested that when shear forces act on a polymeric system, the macromolecules move relative to each other. This results in the destruction of the topological sites forming the structural net. New sites may be formed as a result of Brownian motion, but their lifetimes become shorter and shorter with growing shear forces. The polymer net is gradually destroyed with a decrease in viscosity. However, the situation may be described more adequately by postulating the existence of a spatial net of hydrogen bonds in solution which is destroyed under the action of the shear deformation. This leads to the release of liquid immobilized by the polymer.

It is likely that changes in the rheological properties of solutions of rigid chain polymers with different structures are due to anisotropy of viscosity. This would lead to a dependence of the internal friction on the orientation of long molecular axes along the direction of flow [16]. For systems with isotropic properties such orientation is insignificant, but in the case of anisotropic systems, it predetermines all the peculiarities of the rheological behaviour. The rheological behaviour is largely dependent on the supramolecular structure and type of anisotropic disclinations of the liquid crystal system.

Theoretical predictions of the dependence of viscosity on concentration for both isotropic and anisotropic solutions of rigid chain polymers have been discussed by Doi [17].

The dependence of viscosity on the degree of ordering is a reason for the existence of a maximum in the concentration dependence of viscosity under conditions of shear. The physical reason may be the presence of cooperative orientation processes in LC preparations causing the total friction coefficient to decrease.

Polymer type and solvent nature contribute to the character of the concentration dependence of viscosity at different shear rates, but the general form remains similar to that shown in Figure 6.2 [16]. There is an interrelation between the character of the concentration dependence of viscosity and the phase transitions for aromatic polyamides [18], solutions of polypeptides [19], cellulose derivatives [12], etc. Solutions of cellulose derivatives are characterized by a greater viscosity anomaly when compared with other LC solutions. The extreme character of the variation of viscosity with concentration is the most easily recognized peculiarity of

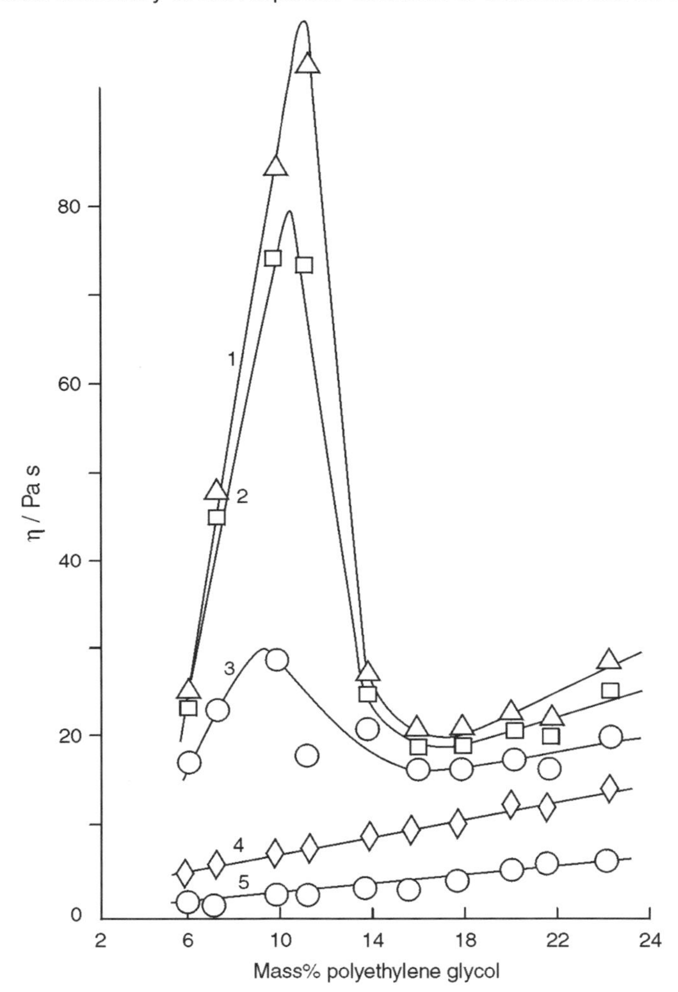

Figure 6.2 Dependence of the viscosity of poly(butyl glutamate) (PBG) on the concentration in *m*-cresol at various shear rates: 1, 0; 2, 10; 3, 100; 4, 1000; 5, 10 000 s^{-1} [16]

the rheological properties of LC polymeric systems. The existence of a viscosity minimum is often a criterion for a liquid-crystalline system when optical characteristics, such as double refraction, cannot be interpreted unambiguously [20]. However a viscosity maximum is not found for cellulose solutions in dimethylacetamide containing LiCl [10, 21]. This system may not undergo an LC transition.

The variation with temperature of the viscosity of LC melts and LC solutions of polymers differs from that of systems with isotropic properties. As in the case of low-molecular-mass liquid crystals, LC polymers are characterized by the existence of extremes of variation of viscosity with temperature under conditions of flow in which the orientation of molecules is not controlled [16]. The interrelation between the character of the temperature dependence and the phase transitions is clearly illustrated for solutions of cellulose derivatives (Figure 6.3).

The rheological behaviour of LC melts of low-molecular-mass substances is similar to that of solutions of rigid-chain polymers. The transition to an LC phase in a state of rest and of flow under shear is a result of structural disclinations.

Fundamental rheological criteria for the identification of liquid-crystalline systems have been proposed [16, 17] on the basis of the experimental data for melts

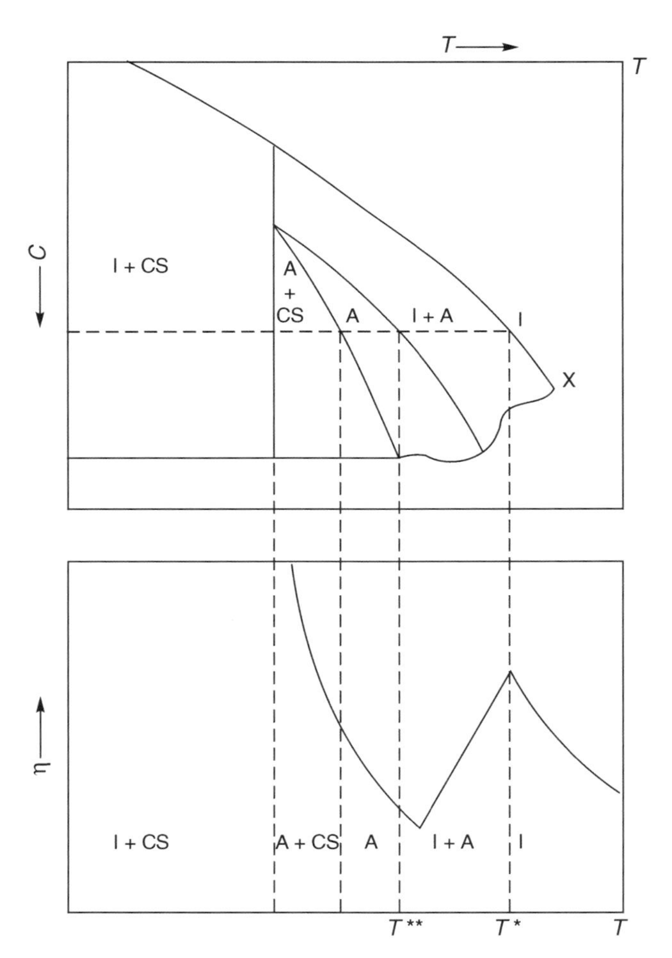

Figure 6.3 Interrelation between the temperature dependence of viscosity and the phase transitions of lyotropic derivatives of cellulose

of comb-like polymers and solutions of rigid-chain synthetic polymers. There is a close relationship between rheological properties and phase transitions. There is special interest in estimating the relationship of rheological indicators of lyotropic liquid crystals with properties of solvents and with the nature of substituents replacing hydroxyl groups in cellulose. Investigation of rheological properties of solutions of cellulose and its derivatives show that there is a wide variety of phenomena associated with the process of flow under the influence of shear stress.

2 INFLUENCE OF POLYMER CONCENTRATION ON THE CONFORMATIONAL STATE OF MACROMOLECULES IN SOLUTION AND ON THE RHEOLOGICAL CHARACTERISTICS

It was assumed in Chapter 2 that intermolecular interactions were absent in dilute solutions when the conformational state of macromolecules was considered. In order to predict isotropic–anisotropic phase transitions in concentrated solutions it is necessary to carry out a detailed investigation of conformational changes due to stronger intermolecular interactions.

This task has already been completed to a large extent for flexible-chain polymers. Theoretical and experimental investigations of solutions of these polymers have indicated certain correlations. The same correlations are also likely to hold for semi-rigid chain polymers, including cellulose and its derivatives. The method of scaling analysis [22] involves a hypothetical transition to concentrated solutions carried out through an extended region of moderately dilute solutions having concentration boundaries C^* and C^{**}. At concentrations at and above C^* there is partial overlap of the macromolecular tangles. A decrease in the statistical dimensions of the molecules ('contraction') and an increase in the number of intermolecular contacts occur as the concentration is increased from C^* to C^{**}.

At concentration C^{**}, the upper boundary of the existence of moderately concentrated solution, 'contraction' of the molecules reaches a stage at which their linear dimensions correspond to those assumed under θ conditions. A further increase in concentration is accompanied by a sharp increase in the number of intermolecular contacts. Under these conditions, a fluctuating net of bonds is formed. This determines the physico-chemical properties of concentrated solutions.

The degree of 'contraction' of the macromolecules depends on the thermodynamic quality of the solvent used. Figure 6.4 shows a diagram of state of solutions of a flexible-chain polymer based on data from theoretical calculations [23]. The diagram illustrates important points. The skeletal rigidity of macromolecules should be much lower in concentrated than in dilute solutions. Solvents used to make concentrated solutions of cellulose must be thermodynamically good. This means that

$$\tau = (T - \theta)/\theta \gg 0$$

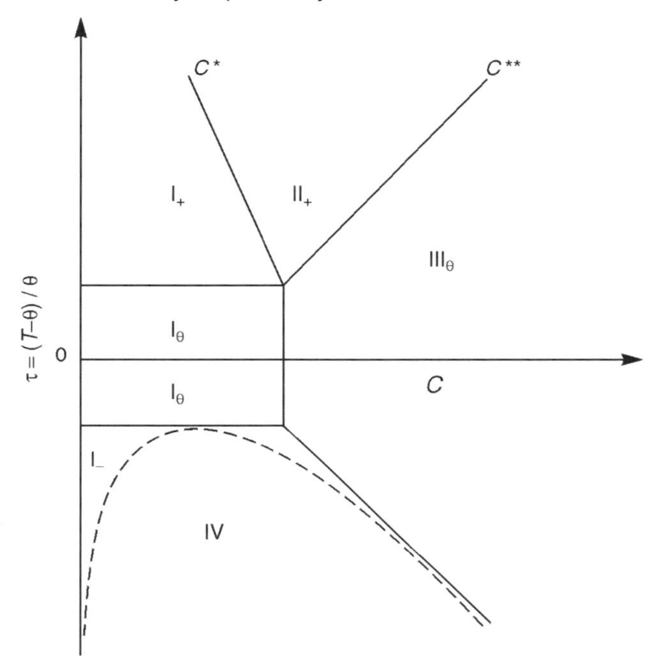

Figure 6.4 Diagram of state of solutions of a flexible chain polymer: I, II and III are regions of dilute, moderately concentrated and very concentrated solutions respectively; IV is the region of phase division; I_θ and III_θ are θ regions ($\alpha = 1$); I_+ and II_+ are regions of positive volumetric effect ($\alpha < 1$) [23]

Baranov *et al.* [24] carried out a hydrodynamic estimation of the perturbing action of the density of the net of intermolecular contacts on the conformation of molecules. The studies showed that the volume occupied by polystyrene molecules dissolved in bromoform decreases with increasing polymer concentration. The higher the mass of the polymer molecules, the sharper is the drop. These findings give qualitative support to theoretical studies.

Information on conformational changes of rigid-chain macromolecules is limited to the data obtained from computer experiments [25–27]. Computer simulation methods show that there is a decrease in equilibrium rigidity of these molecules at the transition from dilute to concentrated solutions. Borisov *et al.* [26] calculated that the branches and frame of a comb-like polymer decrease with increasing concentration beyond the overlap concentration C^*. They also demonstrated that the radial stretching of the arms of a star polymer decreases when its concentration is increased in the moderately dilute range. The arms of one star only begin to penetrate into other stars when the arms reach Gaussian dimensions. Pavlov *et al.* [27] used computer simulation methods to model the behaviour of cellulose nitrate macromolecules in an athermal solution and obtained, in moderately concentrated and very concentrated solutions, macromolecular dimensions that are close to Gaussian with $\langle r^2 \rangle$ equal to $\langle r^2 \rangle_\theta$ (Chapter 5).

3 INFLUENCE OF SHEAR DEFORMATION ON THE RHEOLOGICAL PROPERTIES AND CONFORMATIONAL PARAMETERS OF CONCENTRATED SOLUTIONS OF RIGID-CHAIN POLYMERS

The study of the hydrodynamic properties of polymer solutions by Baranov *et al.* [24] (Section 2) did not include the influence of shear deformation on the conformational state of molecules. The application of an external mechanical field causes deformation of those macromolecules which have pronounced kinetic flexibility. The kinetic flexibility, $\boldsymbol{t}$, characterizes the velocity of the change in chain conformation accompanied by the surmounting of a potential barrier of height U_0 [28]:

$$\boldsymbol{t} \sim \exp(U_0/kT) \tag{6.6}$$

The application of an external mechanical field also causes a decrease in the potential barrier in the direction of the force. This makes conformational transitions associated with the stretching (unwrapping) of the molecular chain in the direction of the force more probable [29]. The unwrapping of the molecular tangles becomes more significant with an increase in the kinetic flexibility of the polymer and an increase in the intensity of the applied field. The increase in the dimensions of a tangle strengthens the hydrodynamic interaction between a molecule and the solvent moving relative to the molecule. This results in additional dissipation of energy by the system.

At the same time, the anisodiametry of a deformed particle increases with shear velocity. The orientation of such a particle in the direction of flow is accompanied by a decrease in the viscosity coefficient of the system. The first effect, the increase in the linear dimensions of a chain molecule due to deformation, results in an increase in the viscosity of the solution. The second, the orientation of the deformed particle during flow, results in a decrease. The second effect tends to predominate as the shear velocity increases. This leads to deviations from Newton's law with an overall decrease in the viscosity of the solution.

A quantitative analysis of the influence of orientation and deformation on viscosity of polymer solutions has been published by Gotlib *et al.* [28]. These are significant only in dilute solutions. The viscous properties of moderately dilute and concentrated solutions are determined by the destruction of the fluctuation net of intermolecular contacts [29]. In the case of these solutions hydrodynamic factors become significant only under conditions in which the fluctuation net is destroyed. This last situation leads to the viscosity tending to become independent of the flow rate with $\eta \to \eta_\tau$ as $\tau \to \infty$, where τ is the shear stress. Sections of such 'quasi-Newtonian' behaviour at high values of tangential tension were found experimentally in flow curves of cellulose nitrate solutions [30]. There is a structural branch between the Newtonian and the quasi-Newtonian sections of a flow curve of such a solution. Viscosity anomalies within the limits of this branch are related to the degree of destruction, α, of the net of intermolecular contacts. The value of α is given by

$$\alpha = (\eta_0 - \eta)/(\eta - \eta_\infty) \tag{6.7}$$

The mechanism of viscosity lowering when the spatial net of intermolecular contacts is destroyed was considered in more detail by Graessley [31]. He introduced the notion of an induction period, θ, that is necessary to form contacts between adjacent macromolecules. The duration of the contacts depends on the shear velocity, the increase of which lowers the probability of formation of intermolecular bonds.

Williams [32] made an attempt to take into account the effects of molecular deformation, the role of which increases while the net of intermolecular contacts is being destroyed. His theory, however, predicts a much more marked viscosity lowering than is observed experimentally [33]. Such a discrepancy may be due to the possibility of the formation of a secondary fluctuation net by the deformed molecules. It was pointed out by Golovko *et al.* [34] that, in systems where molecules are able to interact selectively, deformation strengthens the intermolecular interaction. The increase in the number of contacts then becomes a more dominating factor in comparison with the hydrodynamic effects.

A satisfactory model for the rheological behaviour of polymer solutions, compatible with experimental data, has not yet been developed. This is due to the wide variety of factors which have to be taken into account. Nevertheless, it is possible to assess qualitatively the influence of polymer molecular characteristics, such as molecular mass, on the rheological properties of solutions.

It can be predicted that an increase in molecular mass is accompanied by an increase in the number of intermolecular contacts. The destruction of this net must play an important role in the mechanism of flow. When the macromolecular chain is not long enough to form an effective net of hydrogen bonds, the change in η with increasing τ is likely to be determined by hydrodynamic factors. Because these are small, one would expect the Newtonian character to be retained over a considerable range of shear stress.

These ideas agree with the experimental observation that the dependence of viscosity on M increases when M is above a certain critical value, M_{cr} [35–37]. Values of M are conventionally divided into two groups depending on different values of α in the equation

$$\eta_0 = kM^\alpha \tag{6.8}$$

For the first group different authors give values of α of 1–1.6 [36]. For the second group α is from 3.4 to 3.65 [37]. In concentrated solutions, when the shear stress, τ,

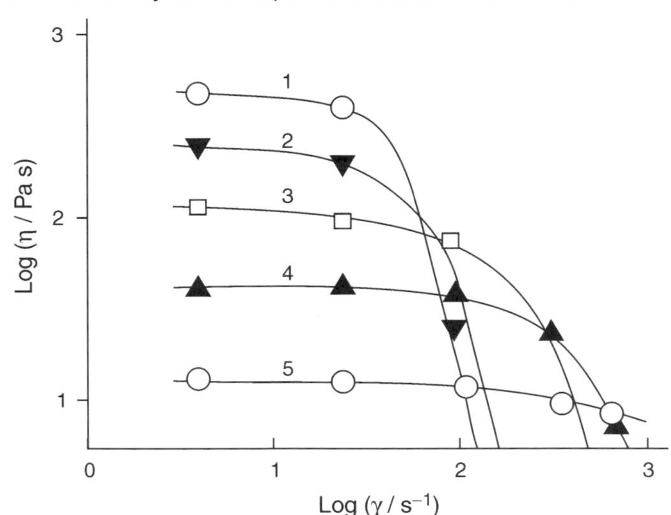

Figure 6.5 Variation of viscosity with shear rate for solutions of (polyterephthalamide *p*-aminobenzhydrazide) ($M = 21\,900$) in DMSO at different polymer concentrations: 1, 6; 2, 8; 3, 10; 4, 12; 5, 14 mass%; $T = 293$ K [39]

tends to infinity and hydrodynamic factors prevail, the viscosity is directly proportional to molecular mass, i.e.

$$\eta = kM$$

over the entire range of molecular mass of the polymer.

The increase in the number of intermolecular contacts with increasing polymer concentration in solution determines the rheological behaviour of solutions. Noticeable deviations from the Newtonian character of flow with increasing concentration are achieved at lower shear rates (Figure 6.5).

4 PECULIARITIES OF THE RHEOLOGICAL BEHAVIOUR OF LYOTROPIC LC SYSTEMS BASED ON CELLULOSE AND ITS DERIVATIVES

The action of a mechanical field in deforming and orienting macromolecules in a polymer solution was discussed above. If a highly oriented state of macromolecules existing in solution can persist when fibres are formed, then these fibres have increased mechanical stability [38, 39]. However relaxation of tensions, which occurs not only when extrudate leaves the filling valve canal but also at the stage of its coagulation, is accompanied by disorientation of macromolecules in the resultant fibre. This has a considerable effect on the characteristics of its stability

[40]. There is much interest in the possibility of spinning fibres from solutions with a high degree of orientation achieved spontaneously due to a high equilibrium rigidity of polymer chains rather than by shear deformation.

The possibility of spontaneous transition to a highly ordered state in concentrated solutions of semi-rigid and rigid-chain polymers, accompanied by an increase in the statistical dimensions of the macromolecules, is confirmed in a series of theoretical [41, 42] and experimental [43–47] studies. Computer simulation of the behaviour of cellulose nitrate in an athermal solution [25] indicated an increase in macromolecular dimensions by approximately 12% when the volume fraction of polymer in solution is between 0.05 and 0.26. At a fraction 0.3 the linear dimensions are only 30% less than the chain contour length (see Chapter 5). According to the theory of the average field of the nematic–isotropic phase transitions [41], it is assumed that there is a spontaneous stretching of the molecular tangles. When the concentration–temperature boundary of the phase transition is reached, the resulting increase in dimensions is predicted to be an order of magnitude for a degree of polymerization of 100, a ratio q/L of 0.1 and an interaction energy between monomeric links of neighbouring molecules of $0.4\,\mathrm{k\,cal\,mol^{-1}}$ (Figure 6.6) (L is the contour length of polymer chain and q is the length of persistence.)

A monograph and two reviews have been published on the rheological behaviour of lyotropic LC polymers [16, 40, 49]. The peculiarities of the viscometric behaviour of spontaneously ordered systems depend on the difference between the viscosity coefficients corresponding to longitudinal and to transversal orientations of anisodiametric particles in flow [50]. This accounts for the marked decrease in the viscosity of an ordered system of such particles oriented along the direction of flow. At the same time, the spontaneous ordering of the macromolecules, owing to their increased equilibrium rigidity in solution, is masked in systems deformed by the stretching and orienting action of a mechanical field. This results in a monotonous increase in viscosity with increase in polymer concentration if the shear velocity is high enough. The situation changes significantly with a decrease in deformation intensity. In this case the ordering of particles is influenced predominantly by their inclination to form a mesophase in solution. This leads to the occurrence of an extreme in the dependence of viscosity on concentration. The occurrence of an extreme is caused by a different degree of disordering of a system as the intensity of deformation is decreased. In solutions in which the polymer concentration is sufficient to form the mesophase, the mutual orientation of the macromolecules is preserved within the limits of the domain structure when the tension is relaxed. This is accompanied by a comparatively small increase in viscosity of the solution.

In the case of flexible macromolecules, qualitatively similar results are caused by the application of an external shear field. Similar results are also caused by increase in the concentration of molecules with a high equilibrium rigidity. Some investigators [51, 52] have tried to induce an LC state in isotropic systems by

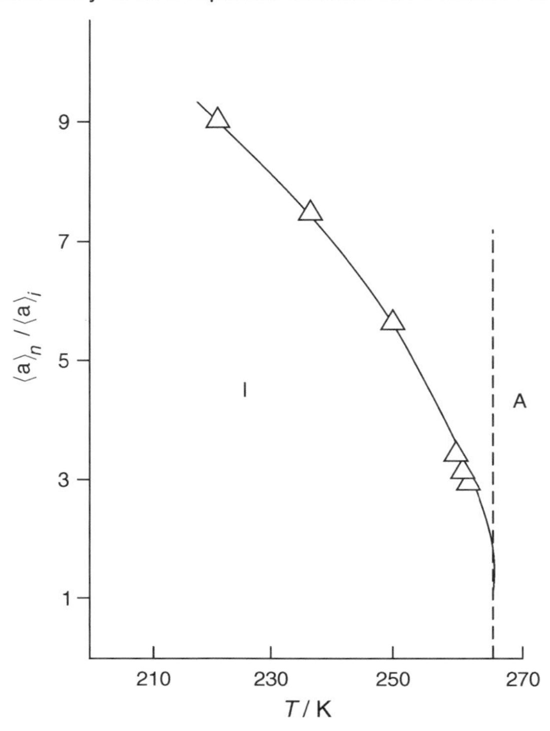

Figure 6.6 Dependence on temperature of the degree of swelling of macromolecules during a nematic–isotropic phase transition at a polymer concentration of 0.8 volume fraction. The dashed line shows the temperature boundary of the isotropic–anisotropic phase transition [41]

the application of a mechanical field. There are theoretical predictions that this should be possible [53, 54]. However, as stated by Papkov and Kulichikin [48], at present there are no direct experimental confirmations of the 'induction' of LC ordering in systems that are not inclined to lyotropic mesomorphism. In the case of solutions of rigid-chain polymers, sometimes deformation facilitates the transition of a system to a highly ordered state [40]. It seems likely that investigators have not taken into account the possibility of an interrelation between the lability of position, the 'pliability' of the temperature–concentration boundary of the phase transition and the kinetic flexibility of macromolecules. The kinetic and equilibrium rigidity do not always change symbiotically [55].

For molecules with high equilibrium and low kinetic rigidity, deformation should favour the realization of the LC state. It is possible that the concept of the kinetic flexibility will eventually be used to explain why, for some rigid-chain polymers such as poly(terephthalamide *n*-aminobenzhydrazide) [40], an increase in shear velocity causes a displacement of the viscosity extreme into the region of lower polymer concentration. This does not occur for some other systems such as polybenzyl-L-glutamate [50].

There is currently great interest in the relationship between the concentration boundary of the isotropic–anisotropic phase transition and the concentration at which the viscosity maximum is reached. Some authors have stressed the similarity between the two concentrations (e.g. Gilbert and Patton [56]). Others consider that the phase transition occurs at a concentration which is lower than that at which the extreme values of viscosity are reached [57]. This latter view is taken by Saito [58], who stated that one should be guided by the position of the maximum in the temperature dependence of the solution viscosity for a more precise estimation of the concentration–temperature conditions for the formation of an anisotropic phase in solution. It is noteworthy that an increase in the intensity of deformation of cellulose solutions in a methylmorpholine-*N*-oxide (MMO)–water mixture and of cellulose triacetate in trifluoroacetic acid does not noticeably influence the position of the maximum in the dependence of viscosity on temperature.

It has been shown by Laivins and Gray [60] that an increase in temperature from 298 to 423 K causes a decrease in the parameter of equilibrium rigidity of acetoxypropylcellulose chains from 150 to 58 Å. Ryskina and Vakulenko [61] detected a similar change when they studied the equilibrium rigidity of cellulose triacetate in acetic acid. At 298 K the parameter was 150 Å and at 333 K it was 90 Å. The concept of a decrease in the chain skeletal rigidity with increasing temperature is often used to explain the displacement of the position of the maximum in the concentration dependence of viscosity into the region of higher polymer concentrations [41]. In particular for CTA solutions in a mixture of trifluoroacetic acid and dichloromethane, Patel and Gilbert [47] found a 50% increase in the concentration corresponding to the maximum value of viscosity when the temperature of the solution increases from 286 to 303 K. Such behaviour is common to all semi-rigid and rigid-chain polymers which form an anisotropic phase in concentrated solutions.

In addition to shear deformation and temperature, the molecular mass of a polymer also has a considerable influence on the position of the extremes in the variation of viscosity with concentration. With increase in M the viscosity maximum occurs at lower polymer concentrations. This may be explained by the influence of equilibrium rigidity on the anisodiametry of molecules and the increase in the chain skeletal rigidity with growing molecular mass. The influence of the molecular mass of a polyglucoside chain on its equilibrium rigidity parameter has been discussed briefly by Ozolinsh and Karkla [62]. Their data indicate a 70% increase in the rigidity of sodium carboxymethylcellulose (NaCMC) chains over a range of low-molecular-mass fractions. Variation of the Kuhn segment value with molecular mass of cellulose derivatives has also been reported by Saito [58].

In some cases a decrease in the skeletal rigidity of polyglucoside chains is observed with increasing molecular mass. Cellulose acetate in acetone behaves in this way. This contrasts with the behaviour of cellulose nitrate in acetone, where there is an increase in rigidity. Indirect evidence for a decrease in equilibrium rigidity of the cellulose chain with an increase in contour length has been reported by Navard and Haudin [63]. They investigated solutions of cellulose *N*-methylmorpholine-*N*-oxide and water mixtures. It was found that the temperature range for the anisotropic state of cellulose became narrower when the degree of polymerization increased from 600 to 900. The contradiction between this observation and the well established tendency of high-molecular-mass fractions of polymers to form a lyotropic LC state can be explained. The author considers that it is unreliable to use anomalously low values of the sensitivity parameter of shear stress, m, in the equation

$$\tau = \eta\gamma^m$$

as the criterion of the formation of an anisotropic phase in solution. Experimental data, that are known to the author, show that it is unsafe to judge the phase state of a deformed system from the degree of deviation from Newtonian character of flow. Some studies indicate that the decrease in the viscosity of anisotropic solutions with increase in shear velocity is less than that with isotropic ones [48]. Experimental data from other investigators indicate the reverse [49].

The existence of a fluidity limit in the curve showing the dependence of viscosity on shear stress, or of the so-called structural branch of textural heterogeneity, is a more reliable indication of anisotropy of a system. Possible interpretations of this phenomenon on the molecular level have been given [48–50].

Various studies have shown how rheological criteria can indicate whether or not a system is likely to be in a liquid-crystal state. The characteristic rheological indicators are a fluidity limit in the flow curves and an extreme character of the concentration and temperature dependence of viscosity. Absence of these indicators casts doubt on earlier identifications of LC ordering in polymer solutions [48].

The author and her colleagues have measured viscosities of solutions of cellulose and its derivatives. The aim has been to correlate rheological properties with the following:

- nature of the solvent;
- chemical character of substituents of OH groups in a cellulose macromolecule;
- molecular mass of polymer;
- polymer concentration;
- temperature;
- shear deformation.

In addition, they have demonstrated an interrelation between rheological properties and phase state. Measurements were made with shear stress (τ) from 1600 to 3000 Pa at 278–338 K. Viscous properties of systems were characterized by the nature of the dependence of viscosity on shear stress and on concentration and temperature.

The rheological behaviour of solutions of HOPC has a series of peculiar features. Of special interest is the comparison of the rheological properties of cellulose, its ethers and esters in common solvents. With these systems it is possible to show the influence of substituents replacing hydroxyl groups in increasing the rigidity of polymer chains. Substituents also affect the development of LC order in the state of rest and flow under conditions of shear deformation [13, 16, 59]. Systems studied included solutions of cellulose, CTA and HOPC in CF_3COOH. When the polymer samples have equal molecular masses of 120×10^3 the flow curves and the variation of viscosity with temperature and with concentration are as shown in Figure 6.7.

Solutions of cellulose in CF_3COOH are characterized by classical flow curves and of variations of viscosity with concentration and temperature. A solution of viscose and also solutions of cotton and wood cellulose with different molecular masses in mixtures of polar aprotic solvents with nitrogen tetroxide behave in a similar way. Flow curves of this type correspond to systems with isotropic properties. This is confirmed by other methods of studying their structures. Cellulose is probably partially trifluoroacetylated in CF_3COOH. Cellulose trifluoroacetate is highly soluble in the acid but is extremely unstable in the presence of moisture. It is hydrolysed when added to water and cellulose is regenerated in a chemically non-modified form. This has been shown by IR spectroscopy and elemental analysis [9].

Cellulose acetates behave in a different way from cellulose when they are dissolved in trifluoroacetic acid. The acetates have a higher solubility, up to 36 mass%. Cellulose has a solubility of 7–10 mass%, depending on the molecular mass of the cellulose. Owing to higher equilibrium rigidity of the polymer chain, the shape of flow curves of CTA solutions in CF_3COOH are very dependent on

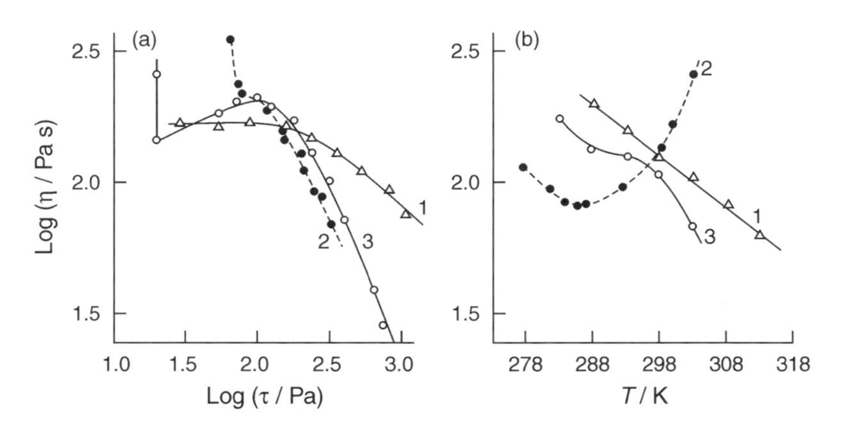

Figure 6.7 Rheological properties of solutions of cellulose, CTA and HOPC: (a) flow curves of solutions of 1, cellulose; 2, CTA; and 3, HOPC in CF_3COOH at 298 K; (b) variation of viscosity with temperature of solutions of 1, cellulose; 2, CTA; and 3, HOPC in CF_3COOH at a shear rate of 1 s^{-1}

concentration. At CTA concentrations above 19 mass%, the flow curves exhibit a series of peculiarities. There is an absence of a range of shear stress in which the viscosity is constant and there is a continuous decrease in viscosity. The flow curves are not convex but are concave in form and the system is inhomogeneous. The concentration and temperature dependence of viscosity under the conditions of shear deformation exhibit extreme character. It has been convincingly shown by the author [64] that such peculiarities are criteria for the existence of polymer systems in an LC state. The rheological evidence that HOPC solutions have anisotropic properties is displayed in Figures 6.1 and 6.7 [9, 21].

The flow curves of HOPC solutions in CF_3COOH are shown in Figure 6.1b. The variation of the viscosity with temperature of HOPC solutions in CF_3COOH with concentrations above 10 mass% passes through extreme values. The rheological properties of solutions show that the tendency towards the LC state in CF_3COOH increases in the series cellulose < HOPC < CTA. The concentration dependence of viscosity shows extreme behaviour.

Addition of chlorinated hydrocarbons to CF_3COOH leads to an increase in the solubility of cellulose and its derivatives without decomposition. It also increases the rigidity of the cellulose chain to 170 Å and hence the tendency towards LC order. CTA and cellulose solutions in CF_3COOH mixtures with $C_2H_4Cl_2$ and other chlorinated hydrocarbons also show extremes in the variation of viscosity with concentration [65] (Figure 6.8).

The degree of substitution of hydroxyl groups of cellulose by the ester groups has a substantial influence on the viscosity of solutions. A good example of this is the rheological behaviour of cellulose acetates with a degree of polymerization of 400 with various degrees of substitution and dissolved in DMSO. The variations in viscosity with change in concentration of solutions of cellulose with degrees of substitution 1.6, 2.4 and 2.9 have been measured at constant shear stress by the author and her colleagues. For equally concentrated solutions, viscosity decreases over the series cellulose triacetate (CTA)–cellulose diacetate (CDA)–cellulose acetate (CA) with decreasing degree of substitution of hydroxyl by acetyl groups [66, 67]. The influence of the solvent is well exemplified by the behaviour of HOPC solutions in a range of non-aqueous solvents.

As was noted earlier (Chapter 1), HOPC possesses thermotropic mesomorphism. It is very soluble in water and in a series of non-aqueous solvents. HOPC (molar substitution $= 3$; $M = 120 \times 10^3$) forms lyotropic LC systems. Figure 6.1 shows the flow curves and Figure 6.9 the variation of viscosity with concentration and temperature of solutions in *N*-methylpyrrolidone. Figure 6.1 shows the transformation of the flow curves as the concentration of HOPC in *N*-methylpyrrolidone increases. It can be seen from the comparison of the viscosities of equally concentrated solutions of HOPC in CF_3COOH, C_2H_5OH and *N*-methylpyrrolidone that η values increase in the series CF_3COOH < C_2H_5OH < *N*-methylpyrrolidone [9]. At the same time, there is a marked similarity in the forms of the flow curves of solutions of HOPC in non-aqueous solvents which have anisotropic properties. The

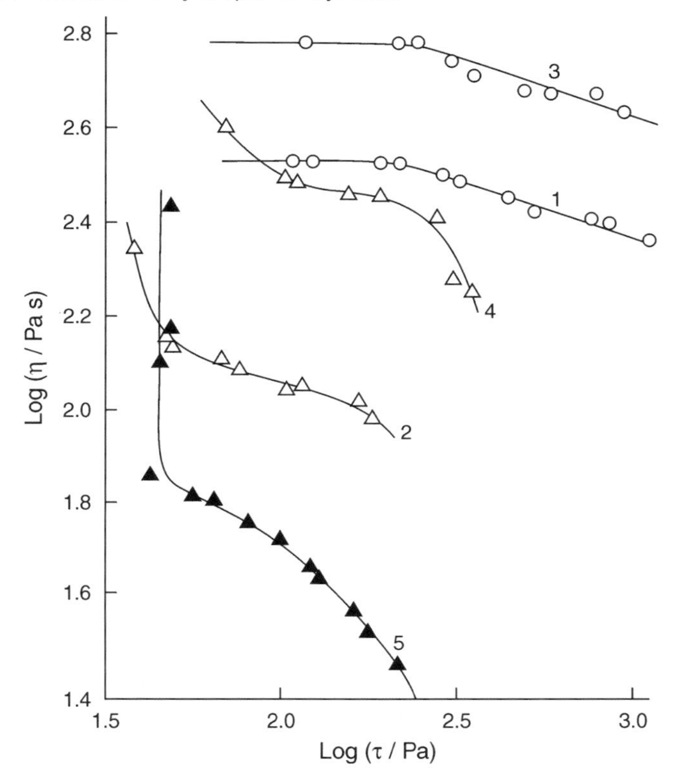

Figure 6.8 Flow curves of CTA solutions in mixed solvent $CF_3COOH{-}CH_2Cl_2$ at $x_{CH_2Cl_2} = 0.57$, $T = 293\,K$ and CTA concentration of 1, 20; 2, 21; 3, 22; 4, 24; 5, 26 mass%

extreme character of the variation of viscosity with concentration and temperature of HOPC solutions in *N*-methylpyrrolidone (Figure 6.9) in combination with the data from polarization-microscopic analysis and circular dichroism has enabled the temperature–concentration boundaries of the LC state to be determined for the first time.

Suto and co-workers [78, 79] measured the transient shear response of liquid crystal-forming hydroxypropylcellulose solution in dimethylacetamide by viscometry. Isotropic solutions had single relaxation and retardation times and liquid crystal solutions had multiple retardation and relaxation times. The stress growth and relaxation behaviour of the liquid-crystalline solutions originated from a change in the liquid crystal domains and depended on shear history. In addition, they investigated simultaneous investigation of band formation and stress relaxation in the liquid crystalline solutions. The higher the shear strain the more rapid was the formation of bands. The speed of formation of bands depended on concentration. In

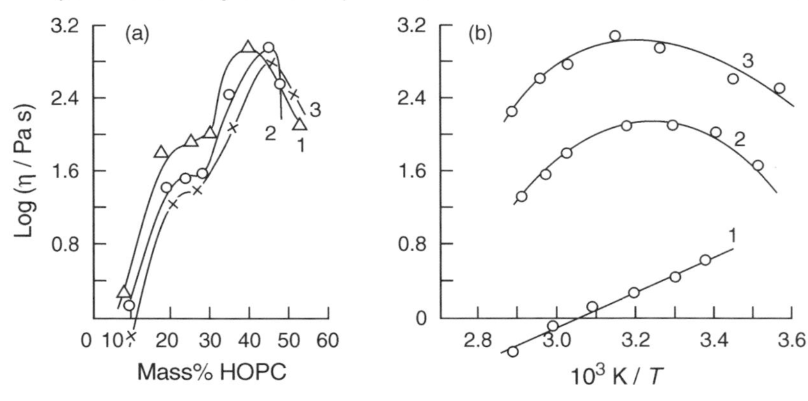

Figure 6.9 Rheological properties of HOPC solutions in *N*-methylpyrrolidone (a) variation of viscosity with concentration at $\gamma = 1.3\ s^{-1}$ at various temperatures: 1, 313; 2, 323; 3, 343 K; (b) variation of viscosity with temperature of solutions of HOPC. 1, 10; 2, 30; 3, 40 mass%

the case of biphasic solutions, an increase in concentration led to an increase in the rate of band formation. In the case of fully liquid-crystalline solutions the reverse was true, with the speed of band formation decreasing with increase in concentration.

Study of CTA solutions in mixtures of CF_3COOH and chlorinated hydrocarbons shows the influence of the nature and composition of non-aqueous solvents on the viscosity of systems with LC order. There are very few additional data in the literature on the comparative investigation of the rheological properties and phase diagrams of non-aqueous solutions of cellulose esters with LC order [65]. Various factors influencing the state of solutions of cellulose and its derivatives have been reviewed [68].

As mentioned above, polymer concentration, non-aqueous solvent composition, temperature and shear deformation exert a considerable influence on the viscosity of CDA solutions in mixtures of CF_3COOH with chlorinated hydrocarbons. Viscosity varies over wide ranges. Flow curves exhibit a variety of forms. The flow curves of anisotropic CTA solutions (degree of substitution, DS = 2.9) in CF_3COOH mixed with chlorinated hydrocarbons are characterized by the absence of the section of Newtonian viscosity. Under certain conditions, the flow curve exhibits a structural branch, which is evidently connected with the textural heterogeneity, and a concave form [64].

Variations of viscosity with change in concentration of CTA solutions in pure CF_3COOH and in mixtures with CH_2Cl_2, CH_3Cl or $C_2H_4Cl_2$ show extreme behaviour. At 0.57 mole fraction of chlorinated hydrocarbon, the viscosity of CTA solutions increases with increasing polymer concentration, passes through a maximum in the region of critical concentrations, decreases sharply on further increase in the polymer concentration and then increases again (Figure 6.10). This

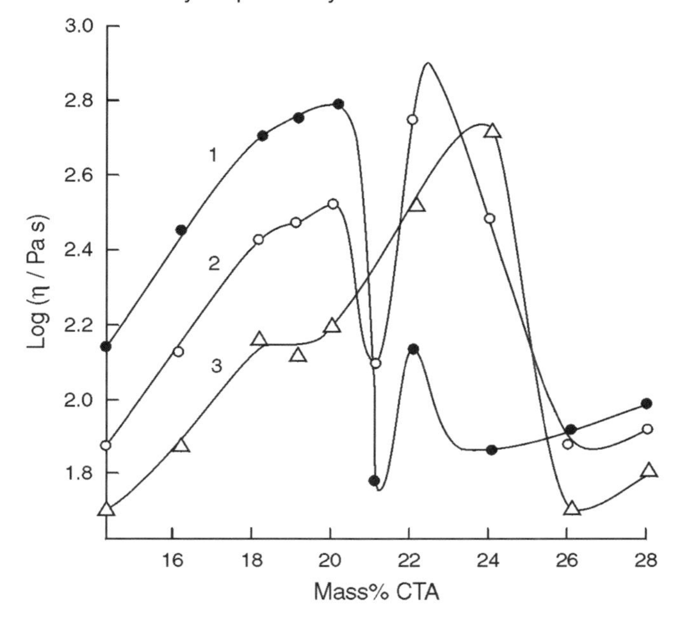

Figure 6.10 Dependence on concentration of the viscosity of CTA solutions in $CF_3COOH-CH_2Cl_2$ mixture ($x_{CH_2Cl_2} = 0.57$ mole fraction) at a shear rate of 1.5 s^{-1}. 1, 278; 2, 293; 3, 303 K

is characteristic of lyotropic and thermotropic liquid-crystalline polymers. The composition of the solvent affects the rheological behaviour characteristic of systems with LC order. When the proportion of chlorinated hydrocarbon in the mixture is high (mole fraction >0.7), the form of the flow curves and the character of the dependence of viscosity on concentration are similar to those for isotropic systems.

Rheological data obtained by the author and her co-workers have made it possible to identify a group of solvents which form solutions which are intermediate between solutions with isotropic properties and those that clearly have LC order. This group of solutions have flow curves which change with temperature and concentration in a way which is characteristic both of solutions with isotropic properties and those with LC properties. This shows that the rheological effects due to the LC state are dependent not only on concentration of CTA but also on temperature.

It also shows that there is a concentration–temperature region in which the phase state is different and characterized by transitional rheological properties. It is likely to be a region in which there is co-existence of isotropic and anisotropic phases. This hypothesis explains qualitatively the behaviour of solutions with transitional rheological properties. With a change in the concentration–temperature conditions, the ratio between co-existing isotropic and anisotropic phases also changes, causing changes in the rheological properties of the solutions [45].

Figure 6.11a shows the positions of the extremes in the concentration dependence of viscosity. These mark the limits of the concentration and temperature of the region. The straight lines going through the extremes distinguish between the three regions with different phases states: regions I and II, where the solutions exhibit isotropic and anisotropic properties, respectively, and region III, where the isotropic and anisotropic phases co-exist [69]. Figures 6.11b and c show the conditions of temperature and concentration for the extreme values of viscosity of CTA solutions in mixed solvents with a CH_2Cl_2 content of 0.57 and 0.67 mole fraction, respectively.

Widening of the double-phase corridor with decreasing temperature is accompanied by the appearance of a ternary point and critical points. At $T < T_t$ and $T > T_c$, the transition from the single-phase isotropic region to the double-phase region containing liquid crystals is accompanied by the appearance of a single maximum of viscosity with change of concentration. Within the $T_t < T < T_c$ interval the increase in polymer concentration results in a repeated change in the phase state of the system. The character of the flow curves and the results of polarization microscopy indicate that, at CTA concentrations below 20 mass%, the solutions possess isotropic properties. A fluidity limit (curve 2 in Figure 6.8) can be clearly observed in the flow curve of the solution at concentration 21 mass%. Absence of the section in which the viscosity is independent of the shear stress is a feature which it shares with flow curves of anisotropic solutions. An increase in CTA concentration by 1% results in a form of flow curve typical of a solution with isotropic properties (curve 3 in Figure 6.8). A further increase in CTA concentration again changes the character of the flow curve. At 24 and 26 mass% of CTA the curve is typical of that of an anisotropic system. Thus, within a narrow concentration range, the character of the rheological behaviour of solutions at $T_t < T < T_c$ changes three times. Here, the maximum values of the viscosity correspond to isotropic solutions when the concentration is 20–22 mass% of CTA. At higher concentrations values of viscosity are lower and the solutions are anisotropic. The experimental facts mentioned above support theoretical predictions of the widening of the double-phase region if the thermodynamic quality of solvent becomes worse [74].

Experimental data from the study of phase equilibria under conditions of shear deformation have enabled the phase diagram of a system consisting of semi-rigid chain polymer (CTA) and solvent (CF_3COOH–CH_2Cl_2) with ternary and critical points to be obtained for the first time. In addition, the interrelation between phase transitions and changes in conformational parameters of cellulose derivatives were studied [78, 79]. The variation of viscosity with change of temperature of CTA solutions in CF_3COOH and its mixtures with chlorinated hydrocarbons also passes through an extreme value (Figures 6.12a and 6.13). The position of the maximum depends on both the polymer concentration and the shear velocity. An increase in

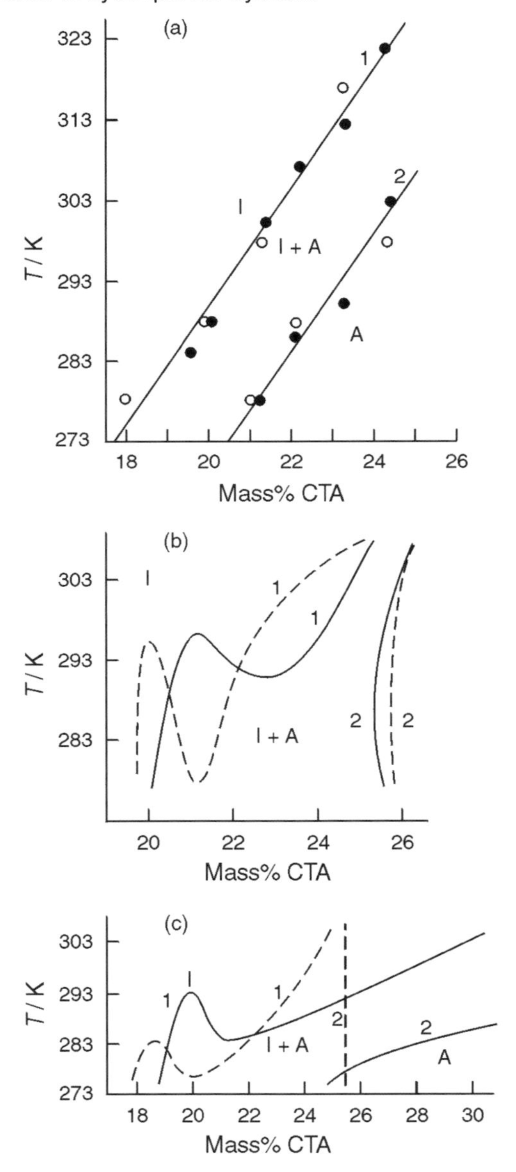

Figure 6.11 Relationships between concentration and temperature for the occurrence of (1) maximum and (2) minimum values of viscosity. (a) CTA in CF_3COOH at shear rate of 1.5 s^{-1}. •, extreme values of the dependence of viscosity on temperature; ○, extreme values of the dependence of viscosity on concentration. (b) CTA in mixtures $CF_3COOH-CH_2Cl_2$ ($x_{CH_2Cl_2} = 0.57$) at a shear rate of 1.5 s^{-1}. (c) CTA in mixtures $CF_3COOH-CH_2Cl_2$ ($x_{CH_2Cl_2} = 0.67$) [68, 69]. In (b) and (c), the continuous lines pass through extreme values of viscosity as the temperature is changed; the dashed lines pass through extreme values of viscosity as the concentration is changed

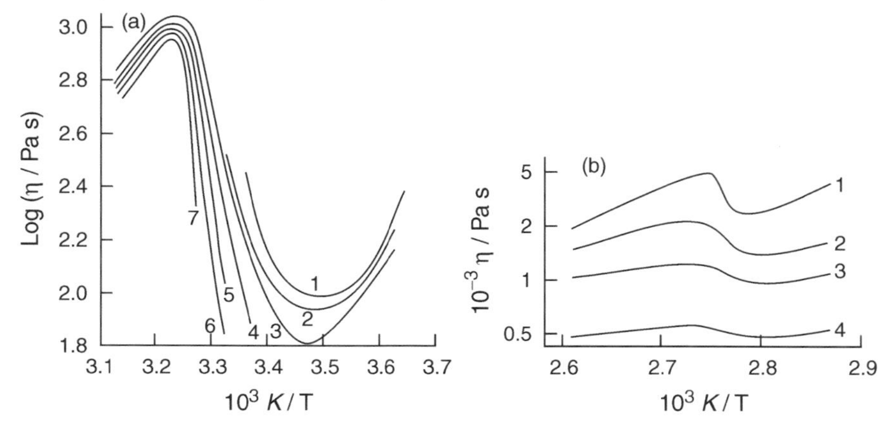

Figure 6.12 Dependence of viscosity on temperature. (a) 22% CTA solution in CF_3COOH at shear stress 1, 1.8; 2, 2.0; 3, 2.2; 4, 2.4; 5, 2.6; 6, 2.8; 7, 3.0 Pa. (b) 24% solution of cellulose (DP = 600) in *N*-methylmorpholine-*N*-oxide–water mixture at shear rates 1, 3.5; 2, 15; 3, 39; 4, 224 s^{-1}

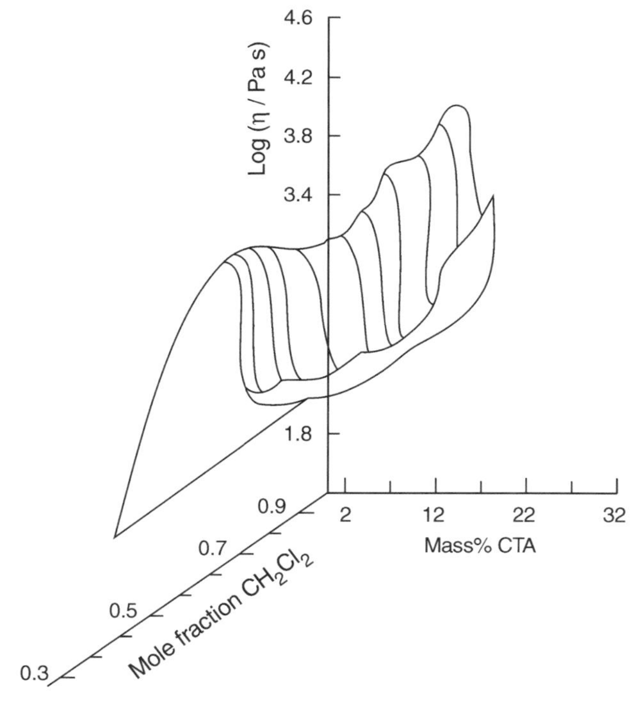

Figure 6.13 Surface of viscosity of the system CTA–CF_3COOH–CH_2Cl_2 at 298 K ($\gamma = 1.5\ s^{-1}$)

shear deformation makes the rheological manifestations of a phase transition to be displaced to higher temperatures. Addition of methylene chloride to trifluoroacetic acid changes the position of the extreme in the variation of viscosity with temperature. The composition of a mixed solvent is an additional factor influencing the formation of the LC state. At a fixed shear deformation the rheological behaviour of LC systems of ethers and esters in non-aqueous solvents can be characterized over the entire range of compositions by a system of geometric surfaces of viscosity within a three-dimensional coordinate system (Figure 6.13) [70, 71].

The dependence of surfaces of viscosity of polymer solutions in individual and mixed non-aqueous solvents on temperature and CTA concentration can be approximated by the general equation

$$\log \eta = \sum_{i=0} \sum_{j=0} A_k \left(\frac{t-25}{25}\right)^{i-j} \left(\frac{C-17}{8}\right)^{j} \tag{6.9}$$

where C is the polymer concentration (mass%), η is the solution viscosity (Pa s) and t is the temperature (° C):

$$k = i(i+1)/(2+j+1)$$

The values of the coefficients A_k, for CTA solutions in $CF_3COOH–CH_2Cl_2$ mixtures with different compositions are given in Table 6.1. When any two parameters, e.g. T and γ, are fixed, the dependence of viscosity on the other parameters of the system ($C_{CTA}, x_{CH_2Cl_2}$) is a surface in ternary coordinates.

Table 6.1 Values of coefficients A_k from Eq. (6.9)[a]

	Mole fraction of CH_2Cl_2 in mixtures with CF_3COOH				
k	0.00	0.13	0.25	0.36	0.47
1	2.743	2.661	2.284	2.190	2.180
2	−0.347	−0.227	−0.163	0.177	0
3	0.731	0	0.226	0	−0.185
5	1.654	1.561	0.660	1.043	0.345
6	−2.219	−2.647	−2.086	−1.086	−0.821
9	0	−0.257	0	−0.324	−0.418
10	−0.380	0.458	0.149	0.361	0.664
11	−0.458	0	0	0	0
13	1.170	0	0	0	0
14	−1.298	−1.445	−0.552	0.735	0.086
15	0.776	1.468	0.972	0.227	−0.113

[a] Note: $A_k = 0$ at $k = 4, 7, 8, 12$ over the whole composition range of the mixed solvents.

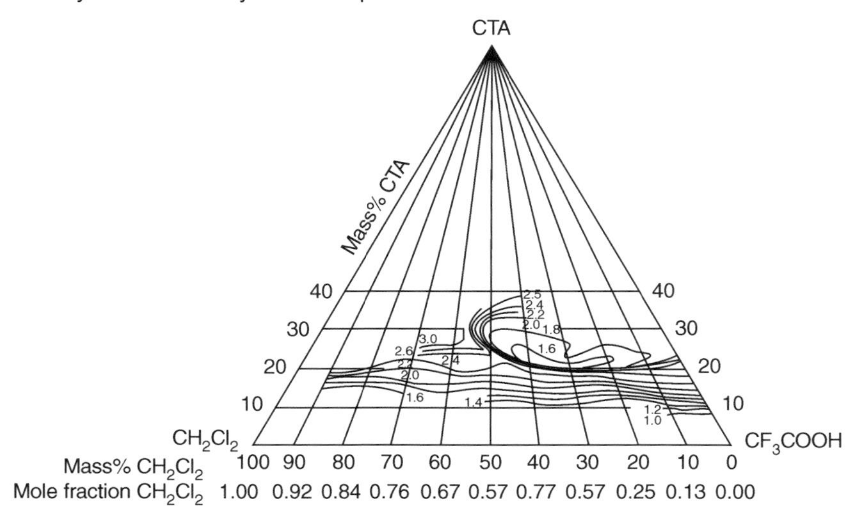

Figure 6.14 Orthogonal projection of the surface of viscosity (isoline diagram) of the system CTA–CF_3COOH–CH_2Cl_2 at 298 K and $\gamma = 1.5\ s^{-1}$

The orthogonal projection of the surface of viscosity on to the plane of the composition triangle shows clearly the influence of polymer concentration and of mixed solvent composition. Figure 6.14 shows a diagram in which there is a series of viscosity isolines at 298 K and $\gamma = 1.5\ s^{-1}$. The density of the isolines indicates the extent to which the viscosity changes with changes in composition of the CTA solution in the mixed solvent CF_3COOH–CH_2Cl_2. The part of the diagram where the density of the isolines is especially high corresponds to the topological region of the surface of viscosity with transitional rheological properties.

An increase in the quantity of dichloromethane in the mixed CH_2Cl_2–CF_3COOH solvent has a negligible effect until the mole fraction of CH_2Cl_2 is > 0.47 before addition of CTA. In pure CF_3COOH the extreme viscosity value occurs when the concentration of polymer is 21 mass%. In a solvent containing 0.47 mole fraction of CH_2Cl_2 it occurs at 19 mass% of polymer. The viscosity reaches a maximum value at a concentration of CTA of 24 mass% when the mole fraction of CH_2Cl_2 in the solvent is 0.57. When the mole fraction of CH_2Cl_2 is > 0.57, the solutions are isotropic at all polymer concentrations which can be achieved.

Extreme values of viscosity when the concentration of a polymer is changed indicate that the LC phase is formed under conditions of shear deformation. There is a good correlation between the concentration–viscosity diagram (Figure 6.14) and the phase diagram (Figure 1.6, Chapter 1) [64, 72, 73]. The effect of an increase in temperature on all solutions of CTA in CH_2Cl_2–CF_3COOH which have been studied is to displace to higher values the CTA concentrations at which the viscosity reaches an extreme value and an LC phase is formed. An increase in temperature also reduces the range of compositions of mixed solvent in which the phase with anisotropic properties is formed. For instance, at 278 K the LC state occurs in CH_2Cl_2–CF_3COOH mixtures at $x_{CH_2Cl_2}$ from 0 to 0.67. At 298 K the range is from 0 to 0.57.

There are two extremes in the change of viscosity with change of concentration of CTA at $x_{CH_2Cl_2} = 0.47, 0.57$ and 0.67 ($\gamma = 0.5\,s^{-1}$) and at $x_{CH_2Cl_2} = 0.57$ and 0.67 ($\gamma = 1.5\,s^{-1}$). Similar results are obtained for CTA solutions in mixtures of CF_3COOH with other chlorinated hydrocarbons and for HOPC solutions in acids and polar aprotic solvents.

The rheological properties of solutions of cellulose and cellulose nitrate differ from those of CTA. Non-aqueous solutions of cellulose with isotropic properties will be considered first. Like solutions of viscose, solutions of cellulose from cotton or wood with different degrees of polymerization are characterized by a classical form of the flow curves. Increasing concentration results in increasing viscosity. The rheological behaviour of systems consisting of cellulose–polar aprotic solvent–nitrogen tetroxide is also typical for systems with isotropic properties.

Significantly different rheological properties are characteristic of non-aqueous solutions of cellulose in mixtures of trifluoroacetic acid with chlorinated hydrocarbons ($C_2H_4Cl_2$, $CHCl_3$), or of methylmorpholine-*N*-oxide (MMO) with dimethyl sulfoxide (DMSO). Rheological measurements have been reported by the author and her co-workers [74, 75]. The way in which viscosity changes with concentration in the region of low shear stress is the most characteristic feature of these solutions. The initial increase in viscosity with increase of cellulose concentration up to a critical concentration, C^*, is accompanied by the appearance of a phase with anisotropic properties. This can be observed with an optical microscope with crossed Polaroids. With further increase in concentration of polymer to 20 mass%, the viscosity decreases but greater concentrations cause the viscosity to increase again. Solutions of cellulose dissolved in a mixture of MMO and DMSO exhibit double refraction due to anisotropy. They also have unusual rheological properties [80, 81]. The systems have been investigated by IR spectroscopy. When the spectrum of the solvent is compared with that of solutions of cellulose it is found that there are changes in the position, intensity and half-widths of the bands, $\nu = 950$–$990, 1000\,cm^{-1}, \nu_{CO} = 1680\,cm^{-1}$ and $\nu_{OH} = 3200$–$3600\,cm^{-1}$. These changes indicate donor–acceptor interactions between cellulose and solvent.

In comparison with DMSO, *N*-oxides possess more marked electron acceptor properties with respect to the hydroxyl groups of various compounds. DMSO does not act as a competitor for the formation of hydrogen bonds. The role of DMSO is that it strengthens polarization of the electron density in NO groups of MMO. This results in the strengthening of the interactions with cellulose hydroxyl groups. The rheological properties of solutions of cellulose in MMO–DMSO mixture, as in cellulose–MMO itself, depend on the structural peculiarities of solvent and the destruction of the net of hydrogen bonds [75–77].

Interparticle interactions, both between pairs of macromolecules and between macromolecules and solvent, considerably influence the equilibrium rigidity of cellulose chains in solutions. It is possible that donor–acceptor complexes of cellulose with solvent are formed.

In the solvent system CF_3COOH–1,2-$C_2H_4Cl_2$ unstable cellulose esters of trifluoroacetic acid, possessing a higher solubility and skeletal rigidity than cellulose, may also be formed. Esterification of cellulose is probably the reason why the physico-chemical properties of solutions of cellulose and of CTA in this mixed solvent are similar.

The stability of cellulose trifluoroacetate increases with the introduction of other substituents. It has been possible to obtain a stable mixed nitric acid–trifluoroacetic acid ester of cellulose, containing 7% of bonded nitrogen and 17% of fluorine, which is not hydrolysed even by hot water. The IR spectra of samples of cellulose, cellulose trifluoroacetate (4% of bonded halogen) and the mixed nitric acid and trifluoroacetic acid ester of cellulose (7% of bonded nitrogen), in the form of the films that were obtained by direct pressing, have been published by Belov *et al.* [9].

The influence of the degree of substitution of hydroxyl groups by the acetyl groups on the magnitude of η is very marked. The viscosity of solutions having the same concentration increases with increase in the degree of substitution.

The rheological behaviour of CTA solutions in the composite solvent MMO–DMSO has been studied. Cellulose and its derivatives are very soluble in this mixture [3, 75]. It was reported by Rozhkova *et al.* [75] that, at low shear stress, there is a section in the flow curves of solutions of CTA in MMO–DMSO in which the viscosity is independent of the shear stress. With increasing shear deformation when $\log \tau > 2$, the viscosity of CTA solutions becomes lower. Evidently, when shear forces of sufficient magnitude act on a polymer system, there is a gradual destruction of the spatial net of hydrogen bonds and viscosity of the solution decreases. CTA solutions in mixtures of MMO and DMSO are characterized by an increase in viscosity with increasing concentration of the cellulose derivative. In this case several distinct sections can be distinguished in curves corresponding to $\eta = f(C)$.

In general, polymeric liquid crystals derived from cellulose are a special class of materials with very unusual viscous properties [78, 79]. When spatial deformation over a wide range of velocities and tensions is being considered, the macromolecules can be represented in flow as orientated linear rods. They are distinct from Newtonian liquids and the Trouton number does not equal three. Despite the fact that for the lyotropic polymers experimental values of normal tensions are usually higher that those of shear stress, they are not elastic, with no deformational strength such as is observed for highly elastic polymers.

The best agreement between theoretical calculations and experimental results is provided by Doi's theory, which uses a two-parameter model, to explain the macroscopic properties of flow. This theory takes into account the rotary diffusion and interaction potential. Its chief limitations are that it does not explain the domain structure which is observed for polymeric liquid crystals or the peculiarities of the transitional behaviour of these systems. In addition, Doi's model is applicable only to nematic liquid crystals, which are modelled as systems with non-aggregated rods. Real systems are very seldom purely nematic, and their mobility and aggregation should be taken into account. However, with a four-parameter model, based on the main theses of Doi's theory at high shear velocities and tensions, it is possible to explain the phase transitions and stick-like orientations assumed by the macromolecules [80, 81].

REFERENCES

1. Klenkova N. I., Khlebosolova E. N. *Cellul. Chem. Technol.* **1977**, *11*, 191–208.
2. Reiner M. *Rheology* Nauka, Moscow **1965**.
3. Tsvetkov V. N., Eskin V. E., Frenkel S. Ya. *Structure of Macromolecules in Solutions* Nauka, Moscow **1964**.
4. Khlebosolova E. N., Golyayev V. G., Klenkova N. I. *Russ. J. Appl. Chem.* **1973**, *46*, 888–93.
5. Conio J., Bianchi E., Ciferri A. *Macromolecules* **1983**, *16*, 1264–70.
6. Kamide K., Saito M. *Polym. J.* **1982**, *14*, 517–26.
7. Senakhov A. V., Kulakov A. I., Repina E. V. *Izv. Vuzov. Tekhnol. Telstil. Prom.* **1986**, *6*, 63–67.
8. Pokrovsky S. A., Myasoedova V. V., Prokofyeva M. V., Khin N. N., Krestov G. A. *Dep. VINITI, N8, 277-B*, **1986**.
9. Belov S. Yu., Strakhova T. B., Telegin F. Yu., Myasoedova V. V., Krestov G. A. *Thermodynamics of Non-Electrolyte Solutions* IKhNR, Ivanovo, **1989**, 51–60.
10. Kulichikhin V. G., Petrova L. V., Khanchich O. A. *Khim. Volokna* **1985**, *2*, 42–44.
11. Strakhova T. B., Nikonova V., Rotenberg I. M., Myasoedova V. V. *Dep. VINITI, N7, 75-B*, **1989**.
12. Petrova L. V., Khanchich O. A., Andreyeva I. N., Dibrova A. K. in *Abstracts of I All-Union Symposium on Liquid Crystal Polymers, Chernogolovka* **1982**, 61.
13. Myasoedova V. V. *Solutions of Non-Electrolytes in Liquids* Nauka, Moscow **1989**, 182–232.
14. Brestkin Yu. V., Volkova A. A., Kutsenko N. I., Meltser Yu. A., Shepelevskii A. A., Frenkel S. Ya. *Vysokomol. Soyedin.* **1986**, *28*, 32–37.
15. Mays J. W. *Polym. Prepr., Am. Chem. Soc. Div. Polym. Chem.* **1987**, *22*, 237–42.
16. Kulichikhin V. G. *Liquid-Crystalline Polymers* Khimiya, Moscow **1988**, 331–72.
17. Doi M. *J. Polym. Sci., Polym. Phys. Ed.* **1981**, *19*, 229–43.
18. Krigbaum W. R. *Polym. Liq. Cryst. Lect. Semin. NY,* **1982**, 275–308.
19. Tsvetkov V. N., Shtennikova I. N., Skazka V. S. *J. Polym. Sci.* **1968**, *16*, 3205–17.
20. Kulichikhin V. G., Golova L. K. *Khim. Dreves.* **1985**, *3*, 9–27.
21. Krestov G. A., Myasoedova V. V., Alexeyeva O. V., Belov S. Yu. *Dokl. Akad. Nauk SSSR* **1987**, *293*, 174–76.
22. Birshtein T. M., Zhulina E. V. *Polymer* **1984**, *25*, 1453–57.
23. Birshtein T. M., Skvortsov A. M., Sariban A. A. *Structure of Polymer Solutions: Scaling and Computer Simulation, Pushchino (Prepr. Nauch. Centr. Biol. Issled.)* **1981**.
24. Baranov V. G., Amribakhshov D. Kh., Brestkin Yu. V. *Vysokomol. Soyedin.* **1987**, *29*, 1190–94.
25. Pletneva S.G., Marchenko G.N., Pavlov A.S. *Dokl. Akad. Nauk SSSR* **1982**, *264*, 109–12.
26. Borisov O. V., Birshtein T. M., Zhulina E. B. *Vysokomol. Soyedin.* **1987**, *29*, 1413–18.

27. Pavlov A. S., Pletneva S. G., Khalatur P. G. *Calculation Methods in Physical Chemistry* KGU, Kalinin **1935**, 87–96.
28. Gotlib Yu. Ya., Darinskii A. A., Svetlov Yu. E. *Physical Kinetics of Macromolecules* Khimiya, Leningrad **1986**.
29. Vinogradov G. V., Malkin A. Ya. *Rheology of Polymers* Khimiya, Moscow **1977**.
30. Philippoff W., Gaskins F. H., Brodnyan J. G. *J. Appl. Phys.* **1957**, *28*, 1118–23.
31. Graessley W. W. *J. Chem. Phys.* **1965**, *43*, 2696–2703.
32. Williams M. C. *AIChE J.* **1966**, *12*, 1064–70.
33. Middleman S. *Flow of Polymers* Mir, Moscow **1971**.
34. Golovko L. I., Rumyantsev L. Yu., Shilov V. V., Kovernik G. P. *Vysokomol. Soyedin.* **1988**, *30*, 2572–77.
35. Chang D.K. *Rheology in Processing of Polymers* Khimiya, Moscow **1979**.
36. Vinogradov G. V., Malkin A. Ya., Kulichikhin V. G. *Success in Rheology of Polymers* Khimiya, Moscow **1970**, 181–205.
37. Pakshver E. A. in *Investigation Methods for Cellulose* (Karlivan V. P., ed.) Zinatne, Riga **1981** 192–204.
38. Ciferri A., Krigbaum W. R., Meyer (eds) *Polymer Liquid Crystals I* Academic Press, New York **1982**.
39. Ciferri A., Valenti B. in *Super-Highly-Modular Polymers* (Ciferri A., Word I. eds) Khimiya, Leningrad **1983**, 151–68.
40. Blumstein A. (ed.) *Liquid-Crystalline Order in Polymers* Mir, Moscow **1974**, 273.
41. Bosch T. A., Maissa P., Sixou P. *Nuovo Chim. D* **1984**, *3*, 95–103.
42. Krigbaum W. R. in *Polymer Liquid Crystals* (Ciferri A., Krigbaum W. R., Meyer R. B., eds) Academic Press, New York **1982**, 275–308.
43. Dayan S., Gilli M., Sixou P. *J. Appl. Polym. Sci.* **1983**, *28*, 1527–34.
44. Conio J., Bruzzone P., Ciferri A. *Polym. J.* **1987**, *19*, 757–68.
45. Suto S. *J. Appl. Polym. Sci.* **1987**, *34*, 1773–76.
46. Myasoedova V. V., Krestov G. A., Alexeyeva O. V., Belov S. Yu. *Russ. J. Appl. Chem.* **1987**, *60*, 2523–26.
47. Patel D. L., Gilbert R. D. *J. Polym. Sci., Polym. Phys. Ed.* **1981**, *19*, 1449–60.
48. Papkov S. F., Kulichikhin V. G. *Liquid-Crystalline State of Polymers* Khimiya, Moscow **1977**.
49. Metner A. B., Prilutski G. M. *J. Rheol.* **1986**, *30*, 661–91.
50. Kiss G., Porter R. S. *J. Polym. Sci., Polym. Phys. Ed.* **1980**, *18*, 361–65.
51. Keller A. in *Super-Highly-Modular Polymers* (Ciferri A., Word I., eds) Khimiya, Leningrad **1983**, 241–66.
52. Capaccio G., Ward J. M. *Polym. Eng. Sci.* **1975**, *15*, 219.
53. De Gennes P. *J. Chem. Phys.* **1974**, *60*, 5030–42.
54. Elyashevich G. K., Frenkel S. Ya. *Orientational Phenomena in Polymer Solutions and Melts* Khimiya, Moscow **1980**, 9–90.
55. Tsvetkov V. N. *Vysokomol. Soyedin.* **1979**, *21*, 2606–23.
56. Gilbert R. D., Patton P. A. *Prog. Polym. Sci.* **1983**, *9*, 115–31.
57. Conio J., Bruzzone P., Ciferri A. *Polym. J.* **1987**, *19*, 757–68.
58. Saito M. *Polym. J.* **1983**, *15*, 213–23.
59. Petropavlovskii G. A., Bochek A. M., Shek V. M. *Khim. Drevesi.* **1987**, *2*, 3–21.
60. Laivins G. V., Gray D. G. *Macromolecules* **1985**, *18*, 1746–52.
61. Ryskina I. I., Vakulenko N. A. *Vysokomol. Soyedin.* **1987**, *29*, 306–12.
62. Ozolinsh R. E., Karkla M. A. *Khim. Drevesi.* **1984**, *3*, 68–70.
63. Navard P., Haudin J. M. in *IUPAC MACRO 83, Bucharest* **1983**, 90.
64. Myasoedova V. V. *Vysokomole. Soyedin. Krat. Soobshch.* **1980**, *30*, 666–68.
65. Belov S. Yu., Myasoedova V. V., Krestov G. A. *Russ. J. Appl. Chem.* **1989**, *62*, 135–39.

66. Rovayev S. S., Myasoedova V. V., Krestov G. A. *Khim. Dreves.* **1987**, *5*, 285–93.
67. Rovayev S. S., Myasoedova V. V., Krestov G. A. *Dep. VINITI, N7, 583-B*, **1986**.
68. Myasoedova V. V., Belov S. Y., Krestov G. A. *Non-Aqueous Solutions in Industry and Technology* Nauka, Moscow **1991**, 34–83.
69. Alexeyeva O. V., Belov S. Yu., Telegin F. Yu., Myasoedova V. V. in *Abstracts of IV All-Union Meeting on Problems of Solvation and Complex-Fomation in Solutions* IKhNR, Ivanovo **1989**, 468.
70. Myasoedova V. V., Krestov G. A., Alexeyeva O. V., Belov S. Yu. *Khim. Dreves.* **1988**, *6*, 15–19.
71. Belov S. Yu. *PhD Thesis* Ivanovo **1989**.
72. Myasoedova V. V., Alexeyeva O. V., Krestov G. A. *Russ. J. Appl. Chem.* **1987**, *60*, 2523–26.
73. Belov S. Yu., Myasoedova V. V., Krestov G. A. *Dokl. Akad. Nauk SSSR* **1989**, *1*, 135–39.
74. Myasoedova V. V., Alexeyeva O. V., Krestov G. A. *Khim. Dreves.* **1987**, *1*, 48–51.
75. Rozhkova O. V., Myasoedova V. V., Krestov G. A. *Vysokomol. Soyedin.* **1987**, *29*, 1599–1602.
76. Rozhkova O. V., Krestov G. A., Myasoedova V. V. *Izv. Vuzov. Khim. khim. Tekhnol.* **1985**, *28*, 67–71.
77. Rozhkova O. V., Myasoedova V. V., Krestov G. A. *Khim. Dreves.* **1984**, *2*, 26–29.
78. Suto S., Tateyama S. *J. Appl. Polym. Sci.* **1994**, *53*, 161–98.
79. Suto S., Kohmoto K., Abe A. *J. Appl. Polym. Sci.* **1994**, *53*, 169–178.
80. Myasoedova V. V., Zaikov G. E. in *New Approach to Polymer Materials* (Zaikov G. E., ed.) Nova Science Publishers, **1995**, 125–51.
81. Myasoedova V. V., Zaikov G. E., *Polym. Year b.* **1997** 3–31.

INDEX